DISORDERED SEMICONDUCTORS

Physics and Applications

DISORDERED SEMICONDUCTORS

Physics and Applications

Anatoly Popov
Moscow Power Engineering Institute (Technical University)

Pan Stanford Publishing

Published by

Pan Stanford Publishing Pte. Ltd.
Penthouse Level, Suntec Tower 3
8 Temasek Boulevard
Singapore 038988

E-mail: editorial@panstanford.com
Web: www.panstanford.com

British Library Cataloguing-in-Publication Data
A catalogue record for this book is available from the British Library.

HANDBOOK OF DISORDERED SEMICONDUCTORS: PHYSICS AND APPLICATIONS
by: Anatoly Popov

ISBN: 978-981-4241-76-2 (Hardcover)
ISBN: 978-981-4241-82-3 (eBook)

Printed in Singapore by Mainland Press Pte Ltd.

Preface

The concepts of translational symmetry and long-range order of atomic arrangement present in single crystals, are the pillars of modern solid state physics. However, single crystals are a very special case of the solid state. Translational symmetry and long-range order of atomic arrangement are absent in most materials. Therefore, the future of solid state physics is in developing a theory that would describe the different states of matter including the theory of single crystals as a special case. Besides the academic interest, materials without long-range order in atomic arrangement are important from the application point of view. Devices based on disordered semiconductors have very wide application. It is difficult to imagine modern life without electrophotographic printers and copiers, liquid crystal displays and monitors, optical discs, cheap solar cells and many other devices based on disordered semiconductors.

However, the peculiarities of disordered semiconductors call for new approaches to the device design as compared to traditional single crystal materials. For example, absence of long range order in structural network allows to apply new methods of properties control without any doping (i.e. structural modification of properties). Moreover, with a correct approach the so called inherent disadvantages of disordered semiconductors can be used to the advantage of the device. For example, low charge carrier mobility allows to combine high resistivity of material with photosensitivity in visible part of spectrum and to create electrophotography printers and copiers. Therefore, the goal of this textbook is to link characteristic features of disordered semiconductors' atomic and electronic structures to the device design process.

The textbook has seven chapters. In the first three chapters we discuss the general concepts of disordered semiconductors, atomic structure of these materials and the structure of energy bands, defects, as well as their electrical, optical and photovoltaic properties. Success of semiconductor device design is defined to a large extent by the possibility of material properties control. Since weak sensitivity to impurities is a distinguishing feature of disordered semiconductors, methods of property control are discussed in chapter four. Use of disordered semiconductors in a device is usually limited to thin films. Such films must have predefined properties and preparation methods must guarantee reproducibility of these properties. These questions are discussed in chapter five. Application of disordered semiconductors in various devices is considered in chapters six and seven. The textbook is based on lectures given by the author to students in the field of electronics and microelectronics at the Moscow Power Engineering Institute

Disordered Semiconductors: Physics and Applications by Anatoly Popov

www.panstanford.com
978-981-4241-76-2

(Technical University) and on the series of lectures given by the author at the Samsung Advanced Institute of Technology.

This textbook is intended for undergraduate-level and graduate-level students in the field of electrical and electronic engineering, nano-electronics, material science and solid state physics. The book is also useful for researchers working in the field of disordered semiconductors.

I am grateful to the Elsevier, Nature Publishing Group, Oxford University Press, Taylor & Francis Group, World Scientific Publishing Co. and Institute of Electrical and Electronics Engineering, Inc. (IEEE) for their permission to reproduce copyright materials. I also wish to thank Dr. Mitsuhiro Kurashige for his aerial photographs not published before. And I am particularly grateful to my sons Dr. Igor Popov and Mr. Alex Popov for their help in preparing the manuscript and helpful discussions.

Anatoly Popov
Moscow Power Engineering Institute

Contents

Chapter One

Introduction

Looking through the history of the solid state electronics, it is not hard to see that the main way of evolution (from the material science point of view) is creation of highly ordered semiconductor materials. First semiconductor devices (rectifiers) have been designed on the base of polycrystalline copper oxides and polycrystalline selenium. Then an era of single crystals began. Since that time one of the most important problems (of semiconductor material science) has been the following: how to bring to perfection the order in crystals. On the other hand, during approximately last 30 years some attention has been focused on the disordered semiconductors. But unlike "conventional" crystalline semiconductors disordered (non-crystalline) materials have no long-range order of arrangement of atoms and translation symmetry. So the question is as follows. The attention to the disordered materials — is this step forward or step back from the point of view of solid state physics and microelectronics?

Let us start with the physics of semiconductors. It is common knowledge that physics of condensed matter mostly succeeded in application to the greatly idealized object — single crystal. On the other hand one can easily see that single crystal is not so common in nature as disordered systems: such as polycrystalline systems, disordered alloys, amorphous semiconductors, glasses and liquids. And it becomes apparent that it is impossible to use even basic ideas of single crystal physics to describe those materials because crystal physics is based on the crystalline latter or on the translation symmetry. As a result we are facing a paradoxical situation when we can not describe the simplest and obvious properties of non-crystalline materials on the basis of "classic" solid state physics.

For example it is known that optical transmission spectrum is determined by the forbidden gap. It means material should be transparent to photons with energy less than the width of the gap and opaque to the photons with energy greater than the width of the gap. The structure of energy bands and forbidden gap is derived from the Brillouin Zone (forbidden gap is outcome of Bregg reflection of electronic waves from the borders of Brillouin Zone). In its turn the concept of the

Disordered Semiconductors: Physics and Applications by Anatoly Popov

www.panstanford.com
978-981-4241-76-2

Brillouin Zone is based on periodic crystal lattice. So far, since there is no periodic crystal lattice in disordered materials it is impossible to use the notions of Bloch function and Brillouin Zone to such an object. In this case we should speak neither of the energy bands and forbidden gap nor of the optical transmission spectrum mentioned above. But then how could we explain why glass in the windows (non-crystalline material) is transparent to light?

Summing up all that has been mentioned it is possible to say that future of the solid state physics is in development of the theory of condensed matter which will include modern single crystal physics as a particular case. The similar point of view has been expressed by Stanford Ovshinsky, president of Energy Conversion Devices Company, USA, and prime mover in the development of amorphous solids and commercial application of these materials. He said: "One does not understand amorphous materials because one understands crystallinity; rather, one can understand crystallinity as a special case when one understands the basic rules of amorphicity" (Ovshinsky, 1991).

It is not an exaggeration to say that it was academic interest in the dielectric properties of a disordered solid in the last 1500 (one thousand and five hundred) years which led to all our electronics and electrical engineering. That solid was amber and its Greek name "electron" has become almost a symbol of our age. So the answer to the first part of the question mentioned above is: disordered solids are to the same extent the past as they are the future of the solid state physics.

The second part of the question is linked to the application of disordered semiconductors in electronics. Disordered semiconductors such as chalcogenide glassy semiconductors, hydrogenated amorphous silicon and its alloys, organic materials are used in various fields of electronics for the fabrication of:

(1) Thin film transistor (TFT) matrices for liquid-crystal display application.
(2) Solar cells on the base of hydrogenated amorphous and microcrystalline silicon and related materials.
(3) Electrophotographic printers and copiers (Xerox-process).
(4) Image sensors and optical recording devices (vidicons, optical disks, phototermoplastic optical recoding media, photoresists and electronic resists).
(5) Threshold and memory switches including phase-change memory devices.
(6) Sensors for medical diagnostics.
(7) Infrared optics elements.

One can say: all or almost all of these devices can be produced on the basis of crystalline semiconductors. So why should we consider use of alternative materials?

The advantages of disordered semiconductors are stemming from:

(1) A number of unique effects that are absent in crystals and that allow creation of devices on the new principles.
(2) Possibility to make devices of large area and low cost.
(3) Possibility to produce active electronics array on the non-crystalline substrate.

1.1 DEFINITION OF DISORDERED STATE

Generally used rule for definition of something is the transition from the general to the particular case. But it is impossible to use this rule in this case because the starting point is physics of crystals. And a crystal is particular case in comparison with disordered matter.

From the point of view of mutual arrangement of atoms there are two cases. The first is crystals. In this case atoms form periodic lattice. In other words, crystals have long-range order of arrangement of atoms or translation symmetry. Diffraction pattern in this case consists of spots (single crystal) or sharp rings (polycrystal). The second case is non-crystalline (disordered) matter. In this case long-range order of atomic arrangement is absent. Loss of long-range order leads to disappearance of sharp reflex on diffraction pattern and to appearance of diffused halos (Fig. 1.1). So it is possible to give following definition: **non-crystalline materials are characterized by the absence of long-range order or translation symmetry.**

The negative character of this definition is not only contrary to the common rules — to define from the general to the particular — but also has very little useful data and sometimes does not give possibility to answer the question: is the sample crystal or non-crystalline material. Indeed, on one hand only ideal single crystal with infinite dimensions has ideal long-range order. Any surface of real crystal is a breach of long-range order. On the other hand the absence of long range order in non-crystalline solids does not mean the total disorder of arrangement of atoms. There is short-range order (nearest neighbors) and medium-range order of atomic arrangement in non-crystalline solids. Ordering in arrangement of atoms stretch for tens and hundreds Angstrom in non-crystalline solids. Hence the answer to our question "is the given sample crystal or non-crystalline solid" depends on length of ordering in atomic arrangement which one assumes to correspond to crystal. In other words, it depends on subjective estimation.

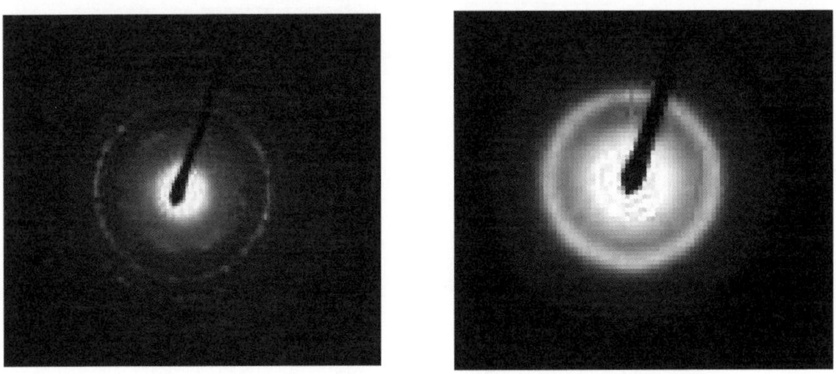

(a) (b)

Figure 1.1. Diffraction pattern of poly-crystalline (a) and non-crystalline (b) materials.

1.2 CLASSIFICATION OF NON-CRYSTALLINE SYSTEMS

Classification is one of natural stages of knowledge. It is necessary to systematize attributes of object prior to study. In other words, any classification is a certain ordering of knowledge about object. Therein is the paradox of this task: to order knowledge of the disordered systems. The problem consists in selection of general characteristics, which could form the basis of classification. And it leads to great variety of classifications. Let us consider the following basic characteristics:

- structural,
- mechanical,
- thermodynamic.

First, structural approach has been considered already — from the structural point of view all materials can be divided into two groups (Fig. 1.2):

- crystals — which have the long-range order of atomic arrangement (or translation symmetry),

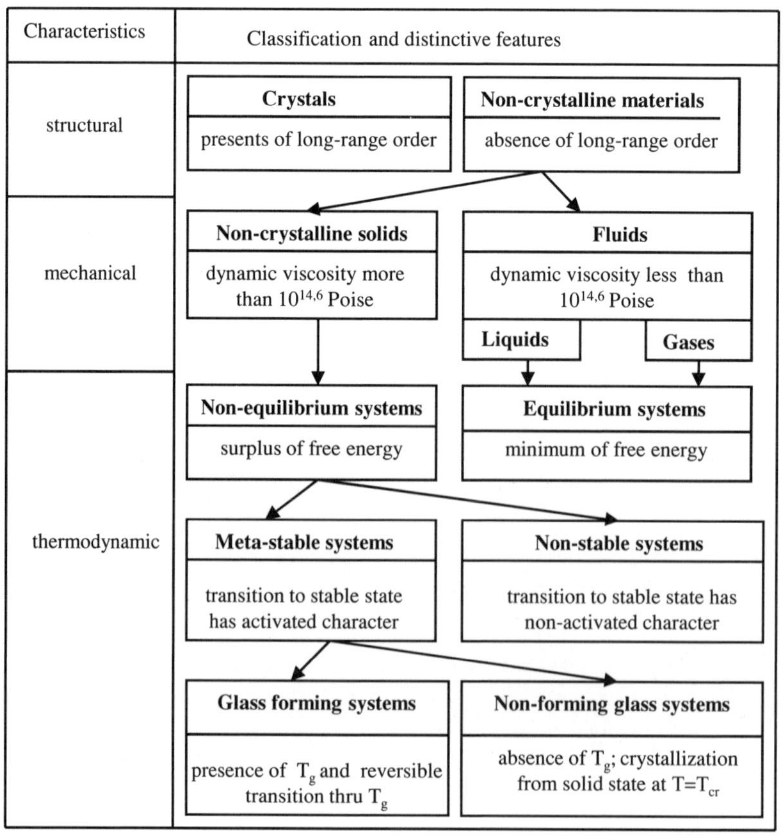

Figure 1.2. Classification of non-crystalline systems.

- non-crystalline materials, at which the long-range order of atomic arrangement is absent.

Disordered or non-crystalline materials from the mechanical point of view can be divided on:

- non-crystalline solids, which dynamic viscosity exceeds value $10^{14.6}$ Poise,
- fluids (liquid, gases), which have dynamic viscosity less than this value.

The boundary point (viscosity equal $10^{14.6}$ Poise) corresponds to relaxation time of system equal to twenty-four hours. Thus, the material is considered as solid, if under action of a small force within twenty-four hours residual deformation is not observed.

In the case of thermodynamic approach energy condition of system is analyzed from the point of view its stability. Stable (equilibrium) system has to have minimum of free energy. Any changes of this system lead to energy increase and therefore to transference it to non-equilibrium state. Crystals, liquids, and gases may be in equilibrium state. Non-crystalline solids are non-equilibrium systems (Fig. 1.2). So there are two characteristic features of non-crystalline solids:

- the absence of long-range order in atomic arrangement and
- non-equilibrium state of system.

Let us examine the non-equilibrium state at greater length. On one hand there is the general rule that any system tends to minimum free energy or to equilibrium state. On the other hand it is well known that glasses remain in non-equilibrium glassy state for hundreds and thousands of years.

So it is necessary to consider the question of stability of non-equilibrium state. As a rule two levels of stability of non-equilibrium materials are named (Fig. 1.2):

- meta-stable states
- non-stable states.

In the case of meta-stable state the system has to cross some intermediate non-stable state with higher energy before it reaches equilibrium state. The other words, there is an energy barrier between meta-stable and stable states. But there is no barrier in the case of transition from non-stable to stable state.

The situation may be illustrated by means of simple mechanical analogue. Let us consider right-angled parallelepiped on the plane. Position on the wide side corresponds to stable state because the centre of gravity is at the minimum height (Fig. 1.3(a)). Position on the narrow side corresponds to meta-stable state. In this case it is necessary to get over energy barrier connected with center of gravity lifting to transfer parallelepiped to stable state (Fig. 1.3(b)). Non-stable state may be presented as pyramid set on its apex (Fig. 1.3(c)). There are no energy barriers in this case.

All transitions in solids are linked to the change of atom positions. That is why these transitions are illustrated by free energy versus configuration diagram

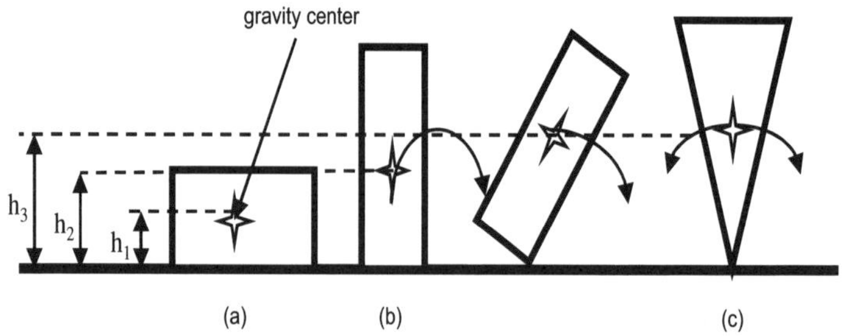

Figure 1.3. Illustration of stable (a) meta-stable (b) and non-stable (c) states (h) is gravity center height.

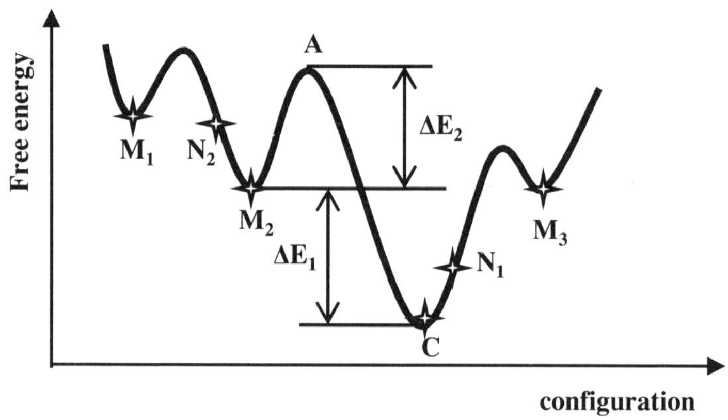

Figure 1.4. Free energy versus configuration diagram.

(Fig. 1.4). Point **C** gives the minimum energy level corresponding to the stable (equilibrium) crystalline state. On the other hand points $\mathbf{M}_1$, $\mathbf{M}_2$, $\mathbf{M}_3$ are local minima corresponding to meta-stable states. Various points of local minima correspond to different meta-stable states which are brought out by different conditions during the formation process or different treatment of material. The difference $\mathbf{E}_M - \mathbf{E}_C = \Delta\mathbf{E}_1$ corresponds to energy surplus of meta-stable state. This energy is exuded during phase transition (for example as crystallization heat). However it is necessary to get over barrier $\Delta\mathbf{E}_2$ for this phase transition to take place (activation energy of crystallization). In the case of thermal excitation the stability of meta-stable states depends on ratio between $\Delta\mathbf{E}_2$ and **kT**.

Let us define the probability of this transition. Transition from meta-stable to stable state means the displacement atoms from point **M** to point **C**. Before atoms reach point **C** they have to get over activated state (point **A**). Equilibrium constant of transition process from point **M** to point **A** is equal to atom concentrations ratio in these states:

$$\mathbf{K} = \mathbf{N}_A / \mathbf{N}_M \tag{1.1}$$

On the other hand equilibrium constant is determined by barrier height ΔE_2:

$$\mathbf{K = exp(-\Delta E_2/kT)} \tag{1.2}$$

In a first approximation it is possible to assume that rate of atom transition from point **M** to point **C** is proportional to concentration of atoms in point **A** (in other words all activated atoms pass into stable state). In this case the process rate may be written as:

$$\mathbf{V = const{\cdot}N_A = const{\cdot}K{\cdot}N_M = const{\cdot}N_M{\cdot}exp(-\Delta E_2/kT)} \tag{1.3}$$

Taking into account that barrier energy may be expressed by enthalpy and entropy changes:

$$\mathbf{\Delta E_2 = \Delta U_2 - T \cdot \Delta S_2} \tag{1.4}$$

we obtain:

$$\begin{aligned} \mathbf{V} &= \mathbf{const{\cdot}N_M \cdot exp\{(-\Delta U_2 + T \cdot \Delta S_2)/kT\}} \\ &= \mathbf{const \cdot N_M \cdot exp(\Delta S_2/k) \cdot exp(-\Delta U_2/kT)} \end{aligned} \tag{1.5}$$

or

$$\mathbf{V = D \cdot exp(-\Delta U_2/kT)} \tag{1.6}$$

where $\mathbf{D = const \cdot N_M{\cdot}exp(\Delta S_2 / k)}$ — temperature independent pre-exponential factor.

Exponential dependence points to strong influence of temperature on process rate. In the case of usual for solids value of $\Delta\mathbf{U}_2$=1.5 – 2.0 eV the value of exponent is equal $10^{-8} - 10^{-10}$ at 1000 K and is equal $10^{-26}10^{-33}$at 300 K. It means that the duration of a process changes from one second at 1000 K to more than 10 billions years at 300 K. So meta-stable states may be rather stable in real conditions.

Typical example of meta-stable systems is glasses. Alongside with the absence of long-range order and thermodynamic non-equilibrium **the distinctive sign of glasses is the glass transition temperature T_g (or glass transition region) and the absence of phase changes in this point**.

Let us consider the dependence of some material property on temperature (Fig. 1.5). When the temperature decreases from their liquid phase, some materials do not crystallize at their melting temperature but they remain liquid (super-cooled) state. As the temperature keeps decreasing, glass transition occurs and below $\mathbf{T_g}$ these materials become solids (glasses). So glassy materials assume the transition from liquid to solid state and back without appreciable changes their properties. From the point of view the free energy versus configuration diagram (Fig. 1.4) it means that transition from solid to liquid state takes place at the temperature below one required to get over barrier $\Delta\mathbf{E}_2$. That is why the system remains in local minimum **M** after cooling.

Alongside glasses there are non-crystalline solids (amorphous silicon for example) that do not have glass transition region but increase the temperature to some critical value ($\mathbf{T_{cr}}$) leads to rapid crystallization from solid non-crystalline

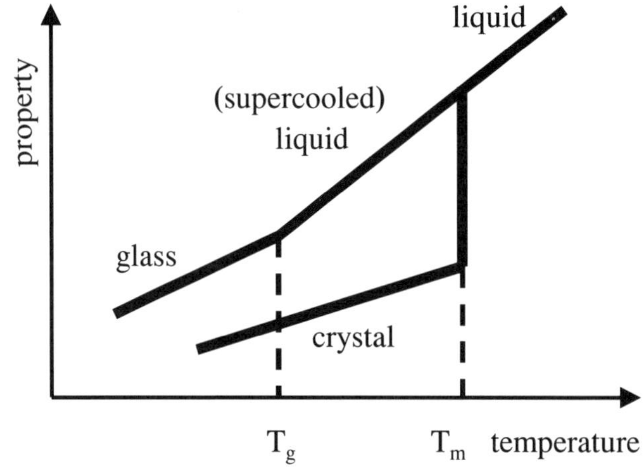

Figure 1.5. Schematic illustration of property verses temperature curve.

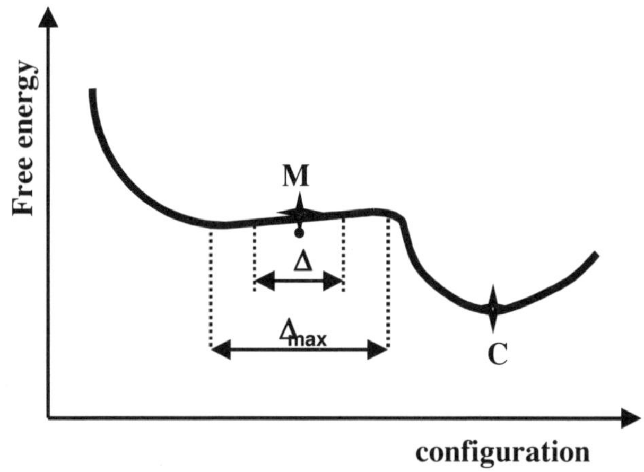

Figure 1.6. Free energy versus configuration diagram for amorphous silicon.

state. In other words, these materials are stable at the temperature less than critical value but at **T** = $\mathbf{T}_{cr}$ irreversible phase transition takes place. The absence of glass transition temperature in meta-stable state is conditioned by features of configuration diagram (Fig. 1.6): small value of barrier $\Delta\mathbf{E}_2$ and the presence of a shelf in the region of point **M**. In this case the system will be in steady state as long as atomic heat oscillation amplitude Δ **is** less than Δ_{max}. Increase of temperature and atomic heat oscillation amplitude up to value $\Delta=\Delta_{max}$ leads to system transition to crystalline state (point **C**).

Non-stable states correspond to points **N** at free energy versus configuration diagram (Fig. 1.4). It should be mentioned that non-stable state may be in vicinity of stable state (point $\mathbf{N}_1$) or in vicinity of meta-stable state (point $\mathbf{N}_2$).

1.3 QUALITATIVE AND QUANTITATIVE CHARACTERISTICS OF GLASS-FORMATION

Glasses are the widespread type of disordered solids. It is possible to prepare material in form of glass only if crystallization is suppressed during cooling of the melt. It means that the possibility of glass preparation depends on both the nature of material and on the rate of cooling. The range of cooling rate is very wide: from 10^{-5} degree per second (change by one degree during 28 hours — fritting of large optical glasses) to 10^6 degrees per second and more (melt-spinning method). In the latter case it is possible to prepare non-crystalline alloys of some metals. But these rates could not be provided for bulk samples. It is possible to prepare only thin films or powder in this case. Usually **material is classified as glass if the cooling rate in the tens of degrees per second is enough to prepare non-crystalline solid.**

Let us consider the following question: why some materials form glasses and others do not? There is no long-range order in the melt. As the temperature keeps decreasing, the rearrangement of atoms to crystal lattice occurs at the melt temperature $\mathbf{T_m}$. The rate of rearrangement depends on mobility of atoms or molecules. Hence the condition of preparing glass is low mobility of atoms or molecules in the temperature range between melt temperature and glass transition temperature. On the other hand the mobility of atoms or molecules is determined by the degree of polymerization. So the first necessary condition for glass formation may be formulated in the following way: material has to consist of high polymer complexes. This condition is necessary but insufficient because any covalent bonded single crystal may be presented as high polymer complex.

The next condition for glass formation comes from the fact that glass formation systems have a predominance of covalent bonds. But this condition is insufficient as well. For example silicon has covalent bonds but it does not form glass. The third condition connected with rigidity of structural network. Glass formation system has to have flexible network. It gives possibility for various atom positions in the matter below melt temperature and glass transition temperature. So the third condition for glass formation consists of the following: the connection of structural complexes has to be only through a single bridge bond. In other words so called swivel bonds must be present in the structure.

Summing up, we may formulate qualitative necessary and sufficient conditions of glass formation:

- material has to consist of high polymer complexes;
- predominant type of chemical bonds has to be covalent bonds;
- swivel (or bridge) bonds have to be present in the structure.

Besides that glass forming ability increases if the number of components of material increase and the glass forming ability decrease in the case of the rise of metallization or ionic degree of chemical bonds.

Let's turn now to quantitative characteristics. There are a lot of theories, concepts, criterions and semi-empirical rules of glass formation (Minaev, 1991). All of them can be divided into three groups: structural — chemical; kinetic; thermodynamic. It is necessary to note that the differences between these groups are rather indistinct. Very often concepts overlap from one group to another. Consider structural — chemical criterion suggested by Dembovsky (1977) as an example. The author attempted to link the glass formation ability to outermost electron shell structure.

Let us consider the atom outermost electron shell structure in the case of elements of IV (silicon) V (arsenic) and VI (selenium). The fourfold coordination of a silicon atom with its neighboring atoms formed by sp^3 hybridized orbitals gives four bonding orbitals and four antibonding orbitals (Fig. 1.7(a)). These orbitals form valence band and conduction band respectively. In the case of selenium the twofold coordination of atom, formed by p-electrons, gives two bonding orbitals,

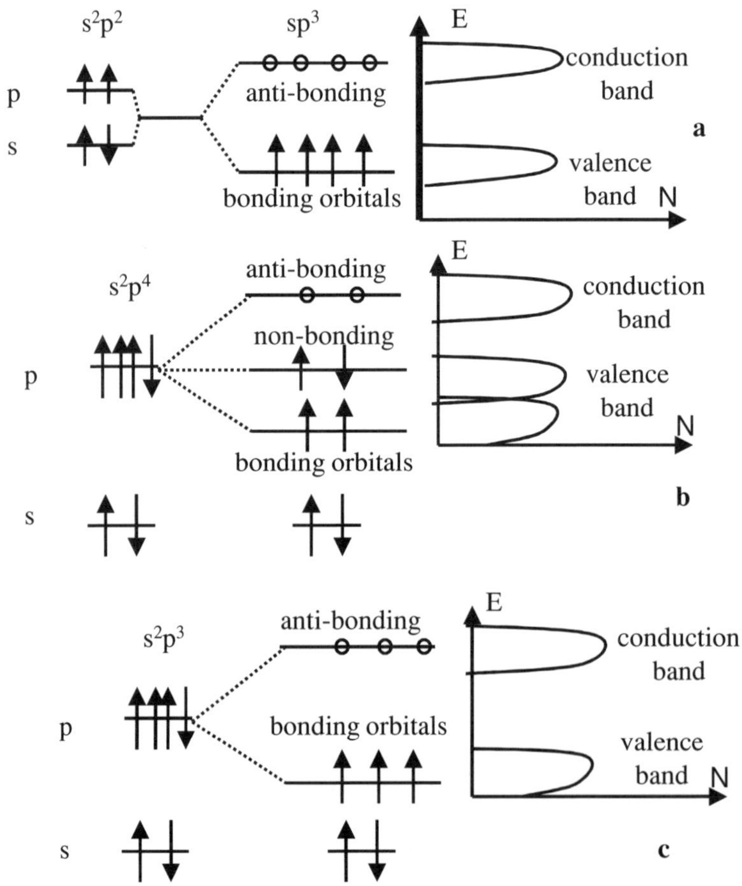

Figure 1.7. Schematic diagrams of the energy levels of atomic orbitals and energy bands for silicon (a), selenium (b) and arsenic (c).

two antibonding orbitals and two nonbonding orbitals (Fig. 1.7(b)). Nonbonding orbitals occupied by lone-pair electrons. Antibonding orbitals form conduction band and nonbonding orbitals form upper part of valence band. Schematic diagrams of the energy levels of arsenic atomic orbitals are given on (Fig. 1.7(c)).

Dembovsky found out that glass formation ability of material has good correlation with the difference between average number of valence electrons (**NVE**) and average coordination number (**N**): **NVE – N**. This difference is equal to the number of electrons on nonbonding orbitals or the number of lone-pair electrons. It was shown that in the case of glass forming materials relative number of long-pair electrons lies in a rather narrow range:

$$\psi = (\mathrm{NVE} - \mathrm{N})/\mathrm{N} = 0.50 \div 0.66 \tag{1.7}$$

For example $As_{40}Se_{60}$ which is good glass-forming material has average number of valence electrons NVE $= 5 \cdot 0.4 + 6 \cdot 0.6 = 5.6$, average coordination number N $= 3 \cdot 0.4 + 2 \cdot 0.6 = 2.4$ and $\psi = 0.57$. It corresponds to the middle of value spectrum in Eqn. (1.7). The value of ψ for selenium equals 0.66. The preparation of glassy selenium is possible only at high quenching rate (cooling of the melt in the water). In the case of silicon $\psi = 0$. This material does not form glass at all.

In the case of criterion ψ less than 0.5 it is not enough swivel (or bridge) bonds in the structure of material. Therefore material has too rigid structural network. If the criterion more than 0.66 the condition of high polymer complexes is broken. Thus the parameter ψ is the quantitative criterion of glass forming ability from structural–chemical point of view.

Table 1.1 Glass forming regions in the systems $A^{IV}B^{VI}$ and $A^{V}B^{VI}$.

System	Chalcogen content, atomic %
As – S	34–49, 55–95
As – Se	40–100
As – Te	42–54
Sb – S	66
Sb – Se	—
Sb – Te	—
Si – S	50–69
Si – Se	80–100
Si – Te	78–80
Ge – S	55–60, 66–90
Ge – Se	75–100
Ge – Te	78–88

The examples of glass formation regions for binary ($A^{IV}B^{VI}$ and $A^{V}B^{VI}$) and triple ($A^{I}B^{V}C^{VI}$) chalcogenide systems are shown at Table 1.1 and Fig. 1.8.

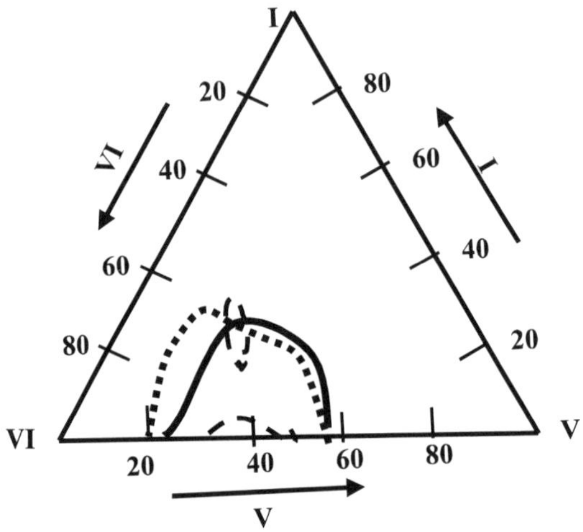

Figure 1.8. Glass formation regions in the systems $A^{I}B^{V}C^{VI}$: Cu–As–Se (solid curve), Ag–As–S (dotted curves) and Ag–As–Te (squares) (Morigaki, 1999).

Chapter Two

Atomic Structure of Disordered Semiconductors

2.1 STRUCTURAL CHARACTERISTICS OF SOLIDS

The atomic structure defines all major properties of both crystalline and non-crystalline substances thus it is important to understand it. At the same time at the XIX International Glass Congress it was pointed out that "commercial as well as research activity is taking place in the absence of anything like a complete understanding of the atomic structure of glass" (Cormack, Du and Zeitler, 2001). Thus, above all, it is necessary to answer the following question: which characteristics of the structure are necessary and sufficient for defining non-crystalline state of solids.

Let us start with the simplest case. From the view point of a structure, the simplest case is ideal single crystal. In order to completely describe its structure it is enough to know the structure of an elementary cell or a short-range order of the arrangement of atoms (Table 2.1) (Popov, 2004). It is necessary to add a defect subsystem for a whole definition of any real single crystal. To describe the structure of polycrystals it should take into consideration morphology of material that is crystal size distribution, crystal texture, formation of spherulites and so on in addition to short-range order and defects. And for non-crystalline solids it should take into consideration four levels of structural characteristics for describing their structure (Table 2.1):

- short-range order of atomic arrangement;
- medium-range order of atomic arrangement;
- morphology;
- defect subsystem.

One can see in the table the amount of characteristics necessary for describing the structure of a substance increases with growth in its complexity. Where one characteristic for describing the structure of an ideal single crystal is sufficient, it is necessary to use four characteristics to describe the structure of non-crystalline solids. It is evident from this that a more complete definition of non-crystalline

Disordered Semiconductors: Physics and Applications by Anatoly Popov

www.panstanford.com
978-981-4241-76-2

Table 2.1 Structural characteristics of solids.

Structure subsystems	Solid states			
	ideal single crystal	real single crystal	polycrystal	non-crystalline solid
short range order	+	+	+	+
defect subsystem	−	+	+	+
morphology	−	−	+	+
medium range order	−	−	−	+

solids in comparison with the definition based on the absence of long-range order of atomic arrangement may be formulated. Non-crystalline solids are materials that require use of the parameters of short and medium range orders of atomic arrangement, morphology and defect subsystem to fully describe their structure. At the same time, as concerns non-crystalline substances, the very terms of short-range order and particularly medium range order of atomic arrangement are presently under discussion.

2.2 SHORT RANGE AND MEDIUM RANGE ORDER

It is known that in absence of long-range order of atomic arrangement in both non-crystalline solids and fluids a certain so called 'local order' remains.

Unlike crystals, where order of atomic arrangement at any level is pre-defined by translation symmetry, understanding of the local order in disordered systems requires specification: what is dimension of fields of local order in atomic arrangement? Which parameters are required and sufficient in order to provide its full description?

When studying elements of order in non-crystalline materials one can choose short-range order of atomic arrangement, determined by the chemical nature of atoms that forms the given substance (valency, bond length, bond angle). As a rule it is assumed that the field of short range order includes the atoms that are the nearest to the atom chosen as a central one and that form the first coordination sphere (atoms 1 and 3 in respect to atom 2 in Fig. 2.1(a), atoms $1, 3, 4, 5$ in respect to atom 2 at Fig. 2.1(b)). The parameters of short-range order are: the number of the nearest neighbour atoms (first coordination number), their type, the distance between them and the central atom (radius of first coordination sphere $-r_1$) (Fig. 2.1), their angle position in respect to the central atom defined by bond angles (valency angles ϕ).

This definition limits short-range order to the first coordination sphere. However, the above mentioned parameters of short-range order define not only the first coordination sphere but at least in part the second one. Thus, the radius of the second coordination sphere r_2 (Fig. 2.1) is determined by the radius of the first

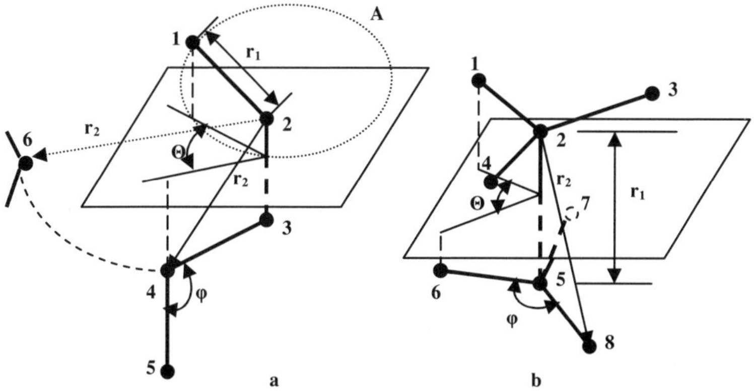

Figure 2.1. Characteristics of atoms' relative positions for linear (a) and tetrahedral structures (r_1 and r_2 are the first and second coordination sphere radii, φ and θ are bond and dihedral angles).

coordination sphere and valency angles:

$$r_2 = 2r_1 \sin\left(\frac{\varphi}{2}\right). \tag{2.1}$$

On the other hand atoms belonging to other molecules may be in the second coordination sphere as well (atom 6 on the Fig. 2.1(a) for example). Positions of these atoms do not connect with short-range order parameters at all.

This contradiction can be resolved if we depart from the geometric parameters of short-range order to the power parameters of interaction among atoms. We can include in the field of short-range order those atoms, respective position of which is defined by the strongest interactions. For semiconductors with predominance of covalent type of chemical bonds the strongest interactions are defined by the parameters of covalent relations (bond-stretching energy and bond-bending energy). Thus, the field of short range order includes atoms of the first coordination sphere as well as those atoms of the second coordination sphere the position of which in respect to the chosen central atom is determined by covalent interaction.

The introduction of the notion of short range order does not allow to describe in full the local order in atomic arrangement observed in disordered systems because it does not explain considerable length of the ordered fields in non-crystalline materials. Experimental proofs of quite long ordered fields triggered introduction of the notion of medium range order in atomic arrangement in non-crystalline materials.

One can consider microcrystalline model of structure of disordered materials as the first attempt to explain the presence of medium range order though the above mentioned models arose historically before the introduction of the term medium range order in atomic arrangement.

According to microcrystalline model suggested by Lebedev (1921) non-crystalline material consists of arbitrarily oriented microcrystalline particles.

Microcrystalline particles separated one from other by disordered layers. Initially it has suggested rather big dimensions of microcrystalline particles (up to 300 Å). Yet later diffraction analysis showed that there are no microcrystalline particles with dimensions larger than 15–20 Å in homogeneous non-crystalline materials. On the other hand Laue patterns from the lower dimensions microcrystalline particles have insufficient contrast to be observed. In other words experimental results indicate absence of microcrystalline particles with dimensions more than 15–20 Å in homogeneous non-crystalline materials but do not prove existence of smaller dimensions microcrystalline particles. However transition to such small dimensions of microcrystalline particles leads to difficulties in interpreting structure on the basis of microcrystalline model. Decrease in dimensions of microcrystalline particles leads to increase of surface atoms number (Fig. 2.2). More than 50% atoms are on the surface of microcrystalline particles if their dimension is less than 50 Å.

In this case disordered layers between microcrystalline particles become more significant than microcrystalline particles for describing structure.

In the case of tetrahedrally bonded semiconductors (amorphous silicon, amorphous germanium and others) models of amorphous clusters (Sadoc and Mosseri, 1982) or amorphons (Grigorovici 1974) have been proposed to explain medium range order. According to these models non-crystalline materials consist of clusters containing a few hundreds of atoms. Atoms are arranged in some regular (but different from crystalline) configuration in the cluster. It leads to decrease of cluster free energy on one hand but it does not give crystal-like diffraction pattern on the other hand. These models were an advance over the microcrystalline model. However, the problem of defining clusters' junctions and their borders was not resolved.

Common shortcoming of these models lays in mechanical dissemination of short range order notions to bigger but geometrically strictly limited regions. It

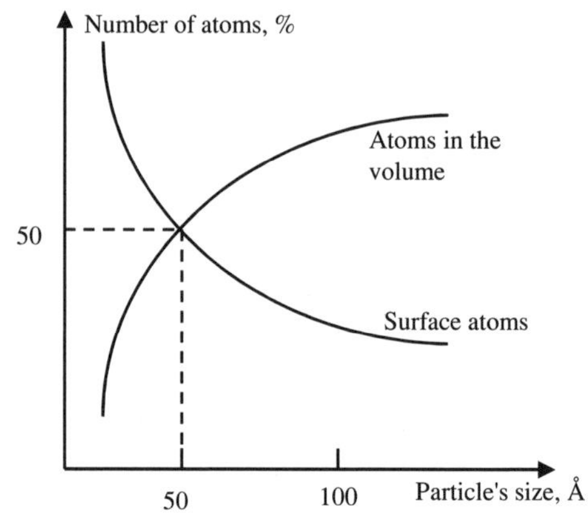

Figure 2.2. Number of atoms in the volume and on the surface of particles dependence on particle's size.

does not allow to solve problems of mutual packing and connection of these regions. Lucovsky (1987) linked medium-range order with the distribution of dihedral angles (dihedral angle is the angle between bonds of nearest neighbour atoms when projected on to a plane perpendicular to the common bond, Fig. 2.1, angle θ) for a distance of approximately ten covalent-bonded atoms. Elliot (1987) divides the field of medium range order into three levels: the field of local medium range order (mutual disposition of neighbouring structural units), medium field of medium-range order (mutual disposition of clusters) and long field of medium-range order linked to spatial order of different fields of a structural network. Voyles *et al.* (2001) offered paracrystalline model for explaining medium range order. But the concrete definitions of this term so far remain open to discussion.

Another approach of defining medium as well as short-range order of atomic arrangement is transition from geometrical characteristics to energy characteristics of mutual atom interaction (Popov, 2004). As mentioned earlier short-range order is determined by the strongest interaction between atoms: bond-stretching energy (V_s) and bond-bending energy (V_b). For example, in case of linear polymer (selenium, sulphur) short range order includes first nearest neighbours (atoms 2 and 4, if atom 3 is considered as the central one, Fig. 2.3) and those atoms of second coordination sphere that are of the same molecule as the central atom (atom 1 as its position in respect to atom 3 is defined by interaction V_s and V_b). Interactions of atoms of the second order include relation of long-pair electrons of atoms that are in the same or different molecules and Van der Waals's interaction between atoms of neighbouring molecules (V_3, V_4, V_{v-v} Fig. 2.3). These interactions determine medium range order of atomic arrangement. Therefore, medium range order is formed by atoms that are partially positioned in the second coordination sphere,

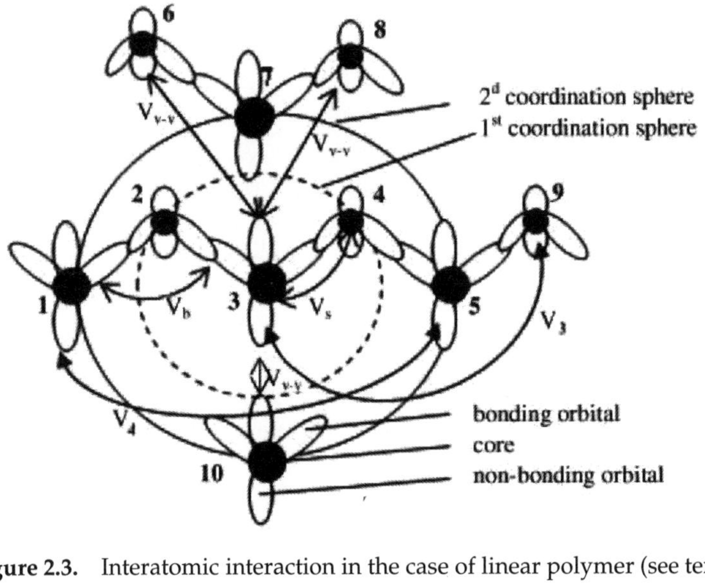

Figure 2.3. Interatomic interaction in the case of linear polymer (see text).

(in case of linear polymer — atoms of neighbour molecules), and the atoms of coordination spheres of higher orders.

So there is no long-range order of atomic arrangement in non-crystalline materials, but there are short and medium range orders. In the case of semiconductor materials with predominance of covalent kind of chemical interaction short-range order is determined by interaction of covalent bonded atoms and includes the first and partially second coordination sphere. Medium range order is determined by interaction of long-pair electrons, Van der Waals's interaction and it is formed by atoms partially positioned in the second coordination sphere and atoms of coordination spheres of higher orders.

2.3 METHODS OF INVESTIGATION OF DISORDERED SYSTEM STRUCTURE

The absence of translation symmetry in disordered systems greatly complicates the task of investigation of their atomic structure as compared with crystals. The direct methods of investigating structure in disordered systems that are similar to crystals are diffraction of short-wave length radiation on the atoms of researched substance and X-ray spectroscopy.

The core of diffraction methods is the registration of spatial picture of intensity of monochromatic radiation coherently dispersed by investigated object, transition from it to distribution of intensity in back space and calculation using Fourier transformation of micro distribution of density of the substance. However if in case of crystals the data obtained in this way gives total information about spatial distribution of atoms in the object, for disordered systems it provides only spherical symmetrical atomic radial distribution function (RDF), that is statistic in character and indicates probability of an atom being positioned at the given distance from the atom chosen as central one.

Three types of monochromatic radiation are used in diffraction techniques: X-ray, electron beam and neutron beam. Interaction with substance is quite different in these cases. X-ray is scattered by electron shells of atoms and contains data about electron density distribution in the substance. Its peaks correspond to positions of atoms. Electron scattering is conditioned by Coulomb interaction and shows the potential distribution of structural network. Potential maximums correspond to atom cores. Neutrons are scattered by delta potential of nuclear force. Interaction intensity of radiation with substance determines requirements to the samples for diffraction analysis. Sample thickness has to be 10^{-5} mm in the case of electron diffractometry, about 1 mm in the case of X-ray diffractometry and up to 10 mm in the case of neutron diffractometry.

Let us consider intensity spatial distribution of coherent scattered monochromatic radiation. The radiation with incident direction $\vec{S}_0$ is scattered in a system of N atoms. Atom positions are described by a set of vectors $\vec{r}$. In this case scattered radiation intensity in the direction $\vec{S}$ is expressed by the following equation:

$$I(s) = \sum_{i=1}^{N}\sum_{j=1}^{N} f_i \cdot f_j \cdot \exp\left[is\left(\vec{r}_i - \vec{r}_j\right)\right] \tag{2.2}$$

where $s = |\vec{S} - \vec{S}_0| = 4\pi \cdot \frac{\sin\Theta}{\lambda}$, λ is wavelength, 2Θ is radiation deflection angle, f is the atomic scattering factor and i is a complex unity. Double sum is taken over all atoms. In the case of $i = j$ vectors of atoms equal as well ($\vec{r}_i = \vec{r}_j$) and exponent equals unity.

In the case of disordered systems vectors of every inter-atomic distance $\vec{r}_{ij}$ may have arbitrary orientation relative to direction $\vec{S}_0$. Therefore spatial averaging of Eqn. (3.2) is necessary. Furthermore the summation in Eqn. (3.2) changes to integration because of a considerable amount of interacting with radiation atoms. Then we obtain

$$\bar{I}(s) = Nf^2 + Nf^2 \int_0^{\infty} 4\pi r^2[\rho(r) - \rho_0] \cdot \frac{\sin(sr)}{sr} dr \tag{2.3}$$

where ρ_0 is the average density of atoms and ρ(r) is radial distribution of atom density. Some manipulation yields

$$s \cdot i(s) = \int_0^{\infty} 4\pi r[\rho(r) - \rho_0] \cdot \sin(sr) dr \tag{2.4}$$

where $i(s)$ is called structure factor

$$i(s) = \frac{\bar{I}(s)}{Nf^2} - 1.$$

Fourier transformation of Eqn. (3.41) provides possibility to calculate radial distribution function from the experimental intensity distribution of scattered radiation:

$$4\pi r^2 \rho(r) = 4\pi r^2 \rho_0 + \frac{2r}{\pi} \int_0^{\infty} s \cdot i(s) \cdot \sin(sr) ds \tag{2.5}$$

In the case of multi-component material Eqn. (3.42) is given by

$$4\pi r^2 \rho(r) \sum_m K_m = 4\pi r^2 \rho_0 \left(\sum_m K_m\right)^2 + \frac{2r}{\pi} \int_0^{\infty} s \cdot i(s) \cdot \sin(sr) ds; \tag{2.6}$$

where m is number of components,

$$i(s) = \frac{I(s) - \sum_m f_m^2}{f_e^2};$$

$$f_e^2 = \frac{\sum_m f_m^2}{\sum_m K_m^2};$$

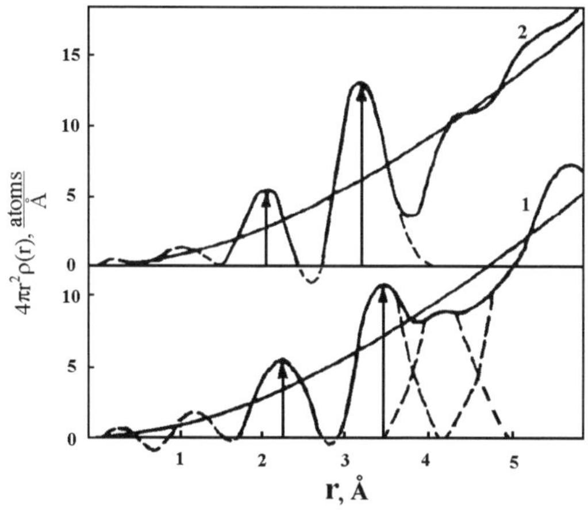

Figure 2.4. Radial distribution function of bulk (1) and thin film (2) a–As_2S_3 (Smorgonskaya and Tsendin, 1996).

$K_m = (Z_m/Z_l)^{0,75}$, f_m and Z_m are atomic scattering factor and atomic number of component m, $Z_l$is atomic number of the lightest component of material and f_e is the value of atomic scattering factor falling at one electron.

Typical RDF of a glassy semiconductor (bulk and film samples of As_2S_3) is given at Fig. 2.4. Disposition and area of the first peak of RDF allows determining the number of nearest neighbours of an atom (first coordination number) and distances to them (radius of first coordination sphere r_1). In case of elementary materials and double compounds of stoichiometric composition in which only different type atoms are connected chemically, disposition of the second peak (r_2) and its halfwidth (δ_2), as a rule give information about average meaning of bond angle ϕ [Eqn. (2.1)] and about range of its changes $\Delta\phi$:

$$\delta_2 = r_1 \Delta\phi \cos\left(\frac{\phi}{2}\right). \quad (2.7)$$

The values of coordination numbers and radiuses of coordination spheres of the third and higher order do not give enough direct information about spatial distribution of atoms beyond the first coordination sphere.

As for binary materials of non-stoichiometric composition and multi-component non-crystalline materials, RDF received from the results of diffraction measuring does not give enough information even for interpretation of short range order parameters. In this case the method of X-ray spectral structural analysis based on extended X-ray absorption fine structure (as a rule, k-absorption) EXAFS is more useful. Analysis of extended X-ray absorption factor allows calculating parameters of first coordination sphere around an atom absorbing X-ray emission quantum. In the case of multi-component material investigation one can get extended X-ray absorption fine structure separately for atoms of each element by changing the energy of X-ray radiation. This allows calculating parameters of first coordination spheres around atoms of each element. But in this case too the information is limited by parameters of short-range order in atomic arrangement.

Thus, diffraction and X-ray spectral analysis methods allow to define parameters of short-range order in atomic arrangement and give some, mainly qualitative, information about structure beyond short-range order but are not sufficient to reproduce spatial disposition of atoms in non-crystalline materials on the base of experimental data.

Another group of structural investigation methods of disordered systems is vibration spectroscopy, including spectroscopy of infrared absorption and Raman scattering, as a rule in the frequency range of 400 to 44 cm^{-1}. In both cases details of received spectra are connected with vibration of atoms and chemical bonds in structural network of materials and therefore give information about forces, operating within structural units. However, infrared and Raman spectra do not duplicate each other, since their rules of selection for transition between oscillatory levels are different.

In case of crystals, harmonic approach is used for decoding spectrums of vibration spectroscopy. In this case definite bands of absorption (also called group frequencies or characteristic lines) in vibration spectrum correspond with single bonds and groups of atoms inside structural units of different chemical compounds.

Transition from crystal to disordered systems further and significantly complicates vibration spectrum calculation due to the loss of long-range order. This is why the method of comparative analysis of obtained spectrums with vibration spectrums of crystals analogous in chemical composition is usually used for interpreting experimental spectrums of non-crystalline materials. The comparison of frequencies of characteristic lines and their intensity in both spectrums allows to draw definite conclusions about the changes of atomic interaction during transition from crystalline state to disordered.

Thus, methods of vibration spectroscopy give information about presence and character of certain bonds and groups of atoms in structural units of investigated substance. At the same time there is no theoretically substantiated analytical correlation that would allow to reliably calculate short inter-atomic distances (not to mention parameters of medium range order) based on the results of vibration spectroscopy. This is due to the fact that inter-atomic distances are only one of many parameters that are the part of kinematic factors of interaction in this complicated vibration spectrum.

In addition to diffraction methods and vibration spectroscopy so-called indirect methods can give certain information about the structure of disordered systems and its change under the influence of different factors. Indirect methods are based on structural dependencies of physical and chemical properties of a substance. As such dependencies exist practically for all material properties and success in using indirect methods is defined by the choice of the most structurally sensitive property of a substance. In case of glassy substances structurally sensitive characteristics include in the first place density, viscosity, solubility, thermal conductivity, heat capacity, sound velocity, refracting index and their temperature dependencies. Since electrical conductivity changes as a rule by several orders of magnitude during transition from glassy to crystalline state its measurement allows to investigate kinetics of such phase transitions.

The analysis of experimental investigation methods of non-crystalline material structure shows that in the best case they give information about short range order in atomic arrangement but none of them separately nor all of them combined provide comprehensive data about medium-range order and therefore they do not allow to reproduce spatial disposition of atoms in disordered systems. At present a simulation of disordered material structure is used for solving this problem.

2.4 SIMULATION OF DISORDERED MATERIAL STRUCTURE

To determine spatial distribution of atoms in disordered systems one can use structure simulation. The results obtained through such structure simulation can then be compared with the same characteristics of the modelled object defined experimentally. There are two methods of creating structural models of non-crystalline materials namely: physical simulation that is creation of model from physical objects (wire, tubes, balls and so on) with further check of model adequacy and correction of atomic coordinates; computer simulation based on some basic atomic distribution and its further transformation until characteristics of model and simulated object are matched. Both approaches are not free from a number of defects. So in case of physical simulation for model construction it is necessary to use the rules of construction based on common conception about the structure of substance (possible mutual disposition of structural units, the meaning of dihedral angles and so on) in addition to the data received from direct experiment (first coordination number, radius of first coordination sphere). Thus, the model is based on both the objective data and subjective perceptions of the author. Some methods of computer simulation are free from such drawbacks. However, insufficient amount of border conditions received from direct experiments causes final distribution of atoms in a model to become only one of multiple possible distributions that match the simulated object.

Physical continuous random network model of amorphous silicon was constructed by Polk (1971). Model contained 440 atoms represented by balls and stems. All atoms (with the exception of atoms on the model surface) have tetrahedral bonds. The author allowed for small bond length fluctuations ($\pm 1\%$) compative to crystal bond length. Mean bond angle was assumed to be equal to bond angle value in crystal ($109^\circ\ 28'$). Random network arose from the absence of limitations on the value of dihendral angle. The model contained two types of cells (Fig. 2.5):

- "straggered" cells with dihendral angle 60° and six-membered rings typical for crystalline silicon,
- "eclipsed" cells with dihendral angle 0° and five-membered rings.

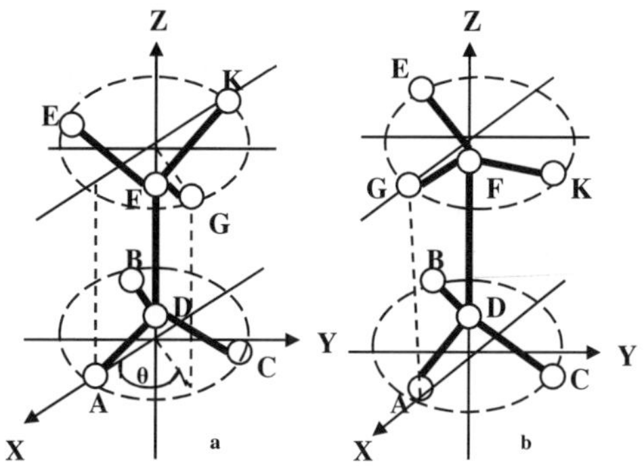

Figure 2.5. "Straggered" (a) and "eclipsed" (b) cells of amorphous silicon.

Formation of "eclipsed" cells resulted in distortion of bond angle. Mean value of bond angle fluctuation was ± 10%. Calculated density of the model was equalled 97 ± 2% in corparison to the actual crystal density. Radial distrbution function of the model adequately reflected experimental radial distrbution function of amorphous silicon.

Structural random network model of 3-fold coordinated amorphous arsenic containing 553 atoms was constructed by Davis and Greaves (1976). The authors tolerated ±20° distortion of the bond angle and placed no restriction on the dihedral andle. One of the objectives during construction was to leave no unsatisfied bonds in the interior of the model. In contrast to the model of tetrahedrally coordinated material it was nessesary to keep apart atoms not directly bonded. Therefore the interlayer space in the model was not allowed to be less than 3 Å. An important feature of the model was that over short distances it was layer like but on a larger scale it was isotropic; that is the layers were so controrted and inter-connected that it was impossible to follow them over distances exceeding 15 Å. This feature may be important for reconciling molecular-like, layer-like and isotropic-like features inferred from certain experimental observations.

Structural model of non-crystalline selenium that has 539 atoms was built by Long and associates, 1976. They have not introduced any limitations on the dihedral angle value in their model. Tetrahedral cells that had two hard covalent bonds and two flexible bonds representing interaction between chain molecules were used for modelling. When designing the model the following limitations were used: presence of large cavities was not permitted, presence of regions with parallel packing of chain molecules similar to microcrystals was not allowed, large deviation of covalent bond lengths and angles as well as break up or rolling of chain molecules within model were excluded. Atomic coordinates of the built model were put into a computer and model strain resulting from deformation of length and angles of covalent bonds as well as intermolecular interaction was minimized. After minimization of strain in the model mean square deviation of length of covalent bonds was 0.89 %, bond angle — 3, 6 % (in respect to 105°). Density of the model was different from the density of non-crystalline selenium by less than 3 %. RDF of the model coincides with experimental RDF of non-crystalline selenium within the limits of first and second peaks.

The main advantage of physical models is their illustrational value. But considerable labour-intensity of designing models with a large amount of atoms caused the principles of constructing physical models to be used mainly for simulating elementary cells of non-crystalline substance when interpreting first peak of radial distribution function. As mentioned in Sec. 2.3 for multi-component and even nonstoichiometric binary non-crystalline materials RDF received from diffraction measurements does not give enough information for definitive interpretation of the elements of short-range order. Constructing physical models of elementary cells, comparing areas and coordinates of first peaks calculated on the basis of the models with similar parameters of experimental RDF allows to determine mutual atomic configuration at least within the first coordination sphere. Coincidence of calculated and experimental parameters of RDF proves correctness of the chosen model to a certain extent.

In computer simulation of atomic structure of glassy materials, molecular dynamics and Monte Carlo methods are widely used. The method of molecular dynamics is used for studying kinetic properties of the matter (for example, for simulating processes of phase transition: crystallization, melting, glass transition) and for constructing models of atomic structure of glassy materials (Cormack, Du and Zeitler, 2001). The method is based on the assumption that movement of atoms can be described by Newton's equations. Force acting on an atom is a sum of vectors $\sum_j^{N-1} \nabla\Phi(ij)$, where N–amount of atoms in system, $\Phi(ij)$–pair-wise interaction potential of atoms (assumed that type of potential is defined), ∇–Hamiltonian. Starting coordinates of atoms are defined pseudo-randomly (that is, with additional condition of prohibition to place two or more atoms at the one point of space) or by the periodic crystalline network configuration. Initial speeds of atoms have random directions and equal absolute meanings chosen so that full kinetic energy of the system is true for the given temperature. When initial coordinates and speeds are set, the atoms are set free one by one and the system begins to approach equilibrium state. The result of simulation is a series of atomic configurations matching different points in time.

Unlike purely deterministic equation of molecular dynamics the Monte Carlo method is a method of numeral calculation in which probability elements are included. The characteristic feature of the method is construction of statically random process where separate states represent various configurations of the examined system that are obtained by random removal of its particles. Every new configuration is accepted or rejected. The criterion for a decision is probability of existence of a new configuration estimated by Boltzman factor $\exp(-\Phi_{Nj}/\mathrm{kT})$ (Φ_{Nj}— potential energy of given configuration) or by the similarity of RDF calculated for the given configuration with experimental RDF.

The main shortcoming of Monte Carlo method is that computer experiment is very time-consuming which leads to limitation of the size of created models. Gradient method of atomic structure simulation was developed to overcome this shortcoming. Unlike Monte Carlo method based on random search for optimum position of atoms in a model, gradient method undertakes targeted search for final disposition of atoms. This allows to reduce the time needed to obtain adequate models of structure significantly, especially in the case of materials where various types of chemical bonds co-exist.

Let us examine gradient method in simulating the structure of glassy selenium (Vasil'eva *et al.*, 1982) as an example. Simulation of structure by gradient method includes several stages. Firstly, the number of atoms in the model is chosen, which allows us to calculate the size of the model while taking into consideration atomic density of the simulated matter. Average density is defined by

$$u = \left(\frac{\rho}{M}\right) m_H, \tag{2.8}$$

where ρ is experimental density of simulated substance, M is atomic weight and m_H = 1,65·10^{-24} g is mass of hydrogen atom. In addition to the size of the model it is necessary to define its form. Cubic or spherical forms are used widely and the choice of a particular form is mainly determined by the method of taking into consideration of the model's finite size. In the case of spherical form various correction factors in calculating characteristics of the model are used to take into account finite size. So, for example, in calculating RDF, errors caused by finite size of a model are compensated by division of RDF by the correction factor defined by

$$D(r) = 1 - 1,5\frac{r}{d} + 0,5\left(\frac{r}{d}\right)^3, \tag{2.9}$$

where d is model's diameter and r is radial distance from the centre of the model. At the same time, not every characteristic of a model can be corrected satisfactorily with the help of correction factors. This is why only the internal part of the model and not its whole volume is used for calculation, so that any atom in it has a normal environment that corresponds to an atom in the volume of material. For all of the above reasons, the examined model of non-crystalline selenium took form of a sphere with diameter of 2.9 nm and included 420 atoms.

After defining size and form of the model the next stage is the creation of initial disposition of atoms. Pseudo-random distribution of atoms in a model is the most acceptable as a starting disposition because it excludes the influence of subjective initial assumptions on the results of simulation. During formation of initial disposition of atoms two limitations should be observed, namely, the number of atoms must equal the pre-defined number of atoms in the model, and the distance between any atoms must not be less than a minimum distance defined by position of the first peak in experimental RDF.

In the third stage of simulation the initial atomic disposition is rearranged in order to match RDF of the model with the experimental RDF. The degree to which they are different is estimated by mean square deviation of these functions:

$$\Delta = \sum_{I=1}^{N} [\mathrm{RDF}_E(I) - \mathrm{RDF}_M(I)]^2, \tag{2.10}$$

where N is number of points, where RDF is compared and $\mathrm{RDF}_E(I)$, $\mathrm{RDF}_M(I)$ are the values of experimental and calculated for the model RDF at point I. The purpose of rearranging atoms in the model is to minimize the value of Δ. When it is necessary to obtain the best possible fit between RDF of the model and experimental RDF at a certain range (for example, in the region of RDF first peak) different weight factors for different ranges of values (I) can be included in Eqn. (2.10):

$$\Delta = K_1 \sum_{I=1}^{N_1} [\mathrm{RDF}_E(I) - \mathrm{RDF}_M(I)]^2 + K_2 \sum_{I=N+1}^{N} [\mathrm{RDF}_E(I) - \mathrm{RDF}_M(I)]^2, \tag{2.11}$$

where K_1 and K_2 are weight factors.

When rearranging initial disposition of atoms using gradient method (as opposed to Monte Carlo method where new atomic coordinates are defined with the help of generation of random numbers) the new atomic disposition corresponding to minimum value of mean square deviation (Δ) is found as follows. Gradient Δ in every direction of coordinate axes is determined for each atom. Based on the obtained values of gradient, the direction in which mean square deviation decreases with the largest speed is chosen. New disposition of the atom in the model that corresponds to minimum Δ in the chosen direction is defined as follows (Fig. 2.6): a cylinder with a radius that equals minimum distance between atoms is built in the direction of the maximum value of mean square deviation gradient. Numbers of atoms, that have their centres covered by the cylinder, are defined. Then, spheres with a radius that equal minimum distance among atoms are built around these atoms. The areas of direct line of maximum gradient Δ covered by these spheres are prohibited for placement of the rearranged atom. After that the values of mean square deviation at the points of overlap between the spheres and the line of maximum gradient Δ are calculated sequentially and an allowed area with the disposition of atom equalling minimum mean square deviation is chosen. In Fig. 2.6 it is the part between points 2 and 3.

Minimum Δ between points 2 and 3 is defined by the method of "golden section". If moving an atom in any direction does not lead to a decrease of mean square deviation, then this atom is left in its initial position. The procedure is repeated sequentially for each atom until desirable similarity between RDF of the model and the simulated object can be achieved.

The fourth stage is the finding of covalent bonded atoms or uniting atoms in molecules. Undertaking this stage after minimizing the value of mean square deviation of radial distribution functions is justified by the fact that rearrangement of free atoms (i.e. atoms that are not linked in molecules) at the previous stage allows to achieve better similarity of experimental and model RDF faster.

Thus, after the fourth stage the model reflects spatial disposition of atoms united into molecules with radial distribution function coinciding with experimental RDF to a given degree of accuracy. However, the obtained spatial atomic disposition is not stable, because energy of such a system does not correspond with local or basic minimum. This is why at the fifth stage of simulation it is necessary to carry out relaxation of the model in order to minimize the total energy of the system, which allows to achieve a realistic atomic disposition.

In the case of tetrahedral bonded systems (silicon, germanium) bond-stretching and bond-bending energy is determined by Keating equation (Keating, 1966)

$$E = \frac{3\alpha}{16d_0^2}\sum_{ij}\left[\left(\vec{R}_i - \vec{R}_j\right)^2 - d_0^2\right]^2 + \frac{3\beta}{8d_0^2}\sum_{ijl}\left[\left(\vec{R}_j - \vec{R}_i\right)\left(\vec{R}_l - \vec{R}_i\right) - \frac{d_0^2}{3}\right]^2 \quad (2.12)$$

where d is the equilibrium bond length, determined by the position of the first peak of the experimental RDF, i, j and l are atom numbers, $\vec{R}_{i(j,l)}$ is vector of atom $i(j,l)$, α and β are bond–stretching and bond-bending constants.

But in the case of two-fold coordinated systems (selenium) the total energy of atoms consists of four components:

- bond–stretching energy between the given atom and its covalently bonded nearest neighbors

$$E_s = 0,5\alpha\sum_{ij}\left(R_{i,j}^2 - d_0^2\right)^2 \quad (2.13)$$

where $R_{i,j} = |\vec{R}_i - \vec{R}_j|$ is the bond length between atoms i and j;
- bond-bending energy

$$E_b = 0,5\beta\sum_{ijl}\left[\left(\vec{R}_{l,i}\vec{R}_{l,j}\right) - c\right]^2 \quad (2.14)$$

where $\vec{R}_{lj} = \vec{R}_l - \vec{R}_j$ and $\vec{R}_{lj} = \vec{R}_l - \vec{R}_j$ are vectors, binding atom l with atoms i and j, c is the constant which is selected in such a way as to make E_b equal zero at the given bond angle φ_0. The meaning of φ_0 is selected as a rule to be equal to a bond angle in the appropriate crystalline material (for selenium φ_0 = 105°);

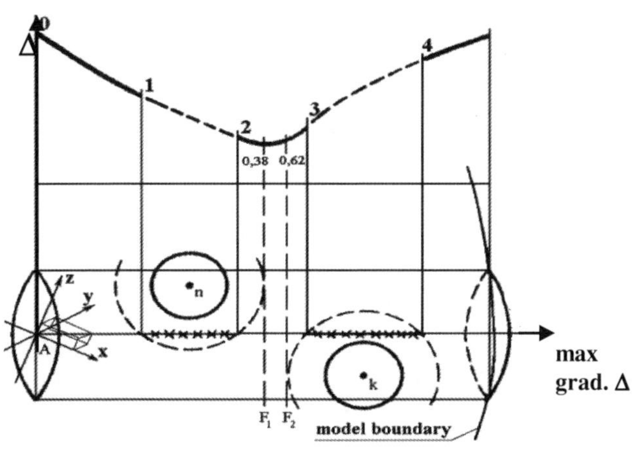

Figure 2.6. Procedure of atom positioning corresponding to mean square deviation (Δ) minimum value finding.

- energy of the van der Waals interaction between the given atom and the atoms which have no covalent bond with it

$$E_V = \sum_i E_{l,i}, \tag{2.15}$$

$$E_{l,i} = \begin{cases} -\frac{A}{R_{l,i}^6} + \frac{B}{R_{l,i}^{12}}, & \text{if } R_{l,i} \leq R_c, \\ 0 & \text{if } R_{l,i} > R_c, \end{cases}$$

  where A and B are the constants defining repulsion and attraction of atoms and selected in such a way that E_{li} equals the minimum value possible with the given distance R_{li}^0 between atoms not directly connected by covalent bonds (in the case of selenium the meaning R_{li}^0 is determined in accordance with the position of the second peak of the experimental RDF and is 3.7 Å), R_c is the maximum distance where the Van der Waals interaction is taken into consideration (as a rule 4–5 Å);
- bond-twisting energy

$$E_d = \gamma \sum_i [(\vec{R}_{ij} \cdot \vec{R}_{jk}) \cdot (\vec{R}_{jk} \cdot \vec{R}_{kl}) - K]^2 \tag{2.16}$$

  where γ is a constant, i, j, k and l are the indexes of sequentially bonded atoms in molecule and K is the constant, the value of which should provide minimum meaning E_d at the given meaning of dihedral angle $\theta = \theta_0$. Value θ_0 is selected as a rule as being equal to the dihedral angle in a certain crystalline modification of the matter (for selenium $\theta_0 = 102°$).

When designing models like these, substantial uncertainty exists in the selection of values for constants α, β and γ. This is why these values are usually selected to provide the required level of similarity between the experimental and model RDF.

Minimization of energy is carried out by sequential movement of each of the atoms in the direction of the quickest decrease of its total energy. This direction is determined by calculating energy gradient of atom, that is, by calculating the force that influences the atom. The search for the position of an atom that would correspond to its minimal total energy in the selected direction is carried out with the help of the algorithm used at the third stage of simulation.

It should be noted that as a rule, the similarity between the model and experimental RDF degrades after the minimization of the system's energy is carried out. In order to enhance the similarity between the experimental and model RDF and to further decrease the total energy of model, the third, fourth and fifth stages of stimulation process are repeated to achieve the desired value of MSD at the minimum of total energy of the model (Fig. 2.7).

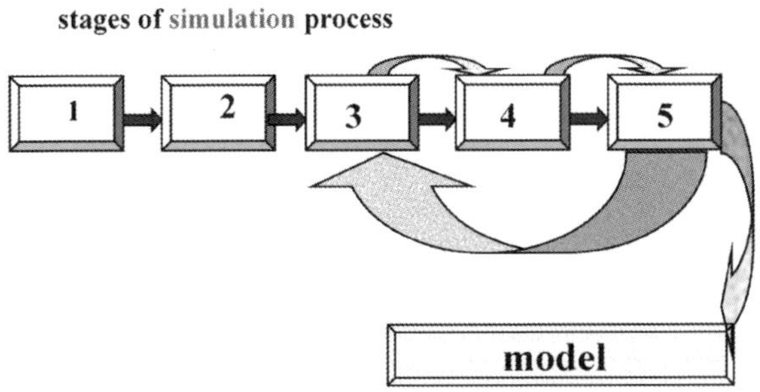

Figure 2.7. Algorithm of atomic structure simulation process (1, 2, 3, 4 and 5 are stages of simulation process).

2.5 RESULTS OF STRUCTURAL RESEARCH OF DISORDERED SEMICONDUCTORS

2.5.1 *Atomic Structure of Non-crystalline Selenium*

Selenium is a member of the VI group of the periodic system. The structure of outermost electron shell is $4s^2p^4$. Hybridization of electron orbitals in selenium is small. Therefore, as a rule, only $p-$electrons form chemical bonds. In elementary selenium two $p-$electrons of each atom form covalent bonds creating molecules in the shape of rings or high polymer chains and the other two $p-$electrons stay in a non-bonding state as lone-pair electrons.

Selenium exists in several allotropic crystalline and non-crystalline forms. Trigonalr selenium, formed by spiral chain molecules Se_n is the thermodynamicly stable form. All other forms of selenium turn into trigonal modification when exposed to thermal treatment.

Crystalline forms of selenium are (Table 2.2, Baratov and Popov, 1990) studied quite well but at the same time a strict classification of allotropic non-crystalline forms is absent. There are at least three allotropic forms of solid non-crystalline

Table 2.2 Allotropic forms of crystalline selenium.

N	Form	Type of molecules	Bond length nm	Bond angle, degrees	Lattice constants, nm		
					a	b	c
1	Trigonal	Spiral chains Se_n	0.233	103.1	0.436	—	0.495
2	α-monoclinc	Rings Se_8	0.232	105.9	0.905	0.908	1.160
3	β-monoclinic	Rings Se_8	0.234	105.5	1.285	0.807	0.931
4	α-cubic	—	0.297	—	0.297	—	—
5	β-cubic	—	0.248	—	0.575	—	—
6	Rombo hedral	Rings Se_6	0.235	101.1	1.136	—	0.442
7	Ortho rombic	—	—	—	2.632	0.688	0.434

selenium: red amorphous, black amorphous and glassy. Red amorphous selenium is produced by chemical restoration, for example, H_2SeO_3, or by sharp quenching of superheated vapour of selenium. It is unstable even at the temperature of about 300 K. The structural models of red amorphous selenium are rather contradictory. However analysis of the obtained data allows to suggest that red amorphous selenium consists of ring-shaped molecules Se_6. At 30–40°C red amorphous selenium is turned into black amorphous modification. This transition has an irreversible endothermic effect that is probably connected with splitting of ring molecules. Information about black amorphous form of selenium is limited by the fact that there is no long-range order in atomic arrangement in this material.

Glassy selenium is the widest spread non-crystalline form of selenium. At the same time information about structure and properties of glassy selenium is greatly different and sometimes even contradictory. It is caused by the fact that glassy state includes various forms of selenium, that differ from each other with ratio, sizes, forms and mutual packing of structural units and therefore they have different properties.

While examining molecular shapes of crystalline forms of selenium (Table 2.2) one can suggest that the same molecules (rings Se_8, rings Se_6, spiral polymer chains Se_n) are present in the glassy matter too. The dependence of glassy selenium properties on the conditions of its preparation is explained by changes in the ratio of different molecules and changes in the level of polymerization of chain molecules. To illustrate this, films of glassy selenium that were prepared in research by Cherkasov and Kreiter, 1974, consisted of only ring molecules Se_8 (amorphous analogue of monoclinic selenium) or of only chain molecules Se_n (amorphous analogue of trigonal selenium) and had as a result widely different properties.

At the same time, absence of long-range order in glassy selenium opens possibility for wider changes to the molecular structure rather than a simple mixing of molecular forms of different crystalline modifications. This is primarily illustrated by the possibility of changing the value or the sign of dihedral angle in molecules and by the possibility of forming defects so characteristic of glassy semiconductors.

The absolute values of dihedral angles for trigonal and monoclinic modifications are close to each other (102° and 101°). The difference is that the sign of dihedral angle in ring molecules of monoclinic form is changed under transition to each next atom so atoms are placed in *cis*-coupling configuration and are closed into rings (Fig. 2.8(a)). The sign of dihedral angle in chain molecules of trigonal form is constant for the whole molecule. In this case atoms are in *trans*-coupling configuration and form an endless spiral chain (Fig. 2.8(b)).

When long-range order is lost, the requirement to have a strictly defined absolute value and sign of dihedral angle is no longer necessary. This is why models of flat zigzag chains (dihedral angle equals zero) (Richter and Breiting, 1971), free rotation chain model in which dihedral angle can have any value (Malaurent and Dixmier, 1977), disordered chains where changes only to sign,

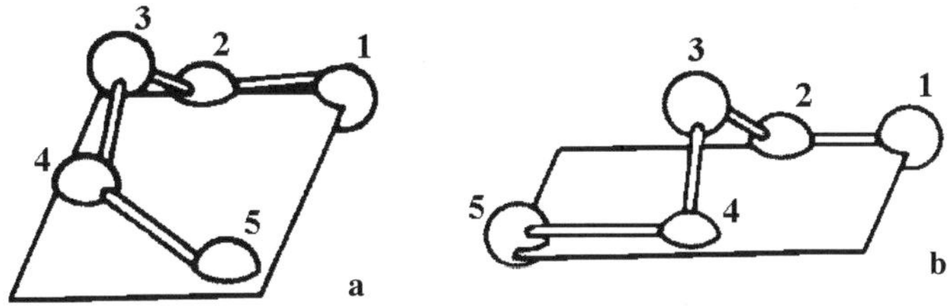

Figure 2.8. Cis- (a) and (b) trans-coupling configuration for molecular bonding in monoclinic and trigonal selenium.

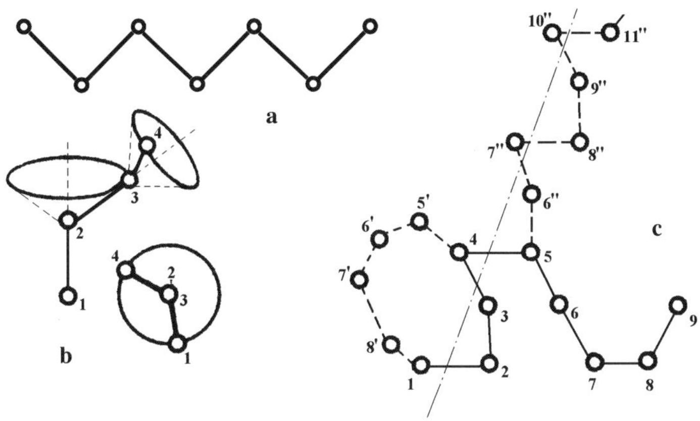

Figure 2.9. "Flat zigzag chain" (a), "free rotation chain" (b) and "disordered chain" (c) models of non-crystalline selenium molecules.

but not to value of dihedral angle are permitted (Lukovsky and Galeener, 1980) were suggested for describing structure of selenium (Fig. 2.9). In the latter case a molecule can have elements of both monoclinic form (rings) and elements of trigonal form (spiral chains).

Other factors that influence glassy selenium structure are quasi-molecular defects and valence alternation pairs (VAP). VAP concentration in selenium can achieve the value comparable with concentration of molecules ($5{\cdot}10^{18}$–$5{\cdot}10^{19}$ cm^{-3}) that must have an effect on the structure of matter due to formation of intermolecular bonds by Se_3^+ centres. The introduction of quasi-molecular defects also leads to formation of intermolecular bonds and to increasing selenium atom coordination (see Chapter 3).

The structure of glassy selenium was widely investigated with the help of diffraction methods and the analysis of extended X-ray absorption of fine structure (EXAFS). The first peak of radial distribution function (RDF) (that is determined by the nearest neighbours in the same molecule) is well isolated from the rest of this curve. The second peak of RDF has a shoulder on the side of large "r". It is caused by the fact that contribution to the second peak is made both by the atoms situated in the molecule (partial coordination number equals two) and the atoms of neighbouring molecules. Decomposition of the second peak into two sub-peaks of Gaussian form under the condition that the area of the first sub-peak corresponds to coordination number two allows to estimate the average value of bonds angle (100°–105°) and the value of deflection from medium meaning (~8°). A number of trigonal form coordination radii (Table 2.2) is absent in the radial distribution function of glassy selenium. An example could be radius 4.36 Å that fits the network constant "*a*" in trigonal form. Absence of appropriate coordination sphere is due to the loss of periodical disposition of molecules. Disposition of the third peak RDF of glassy selenium (4.7–4.9 Å) is close to the value of network constant "*c*" of trigonal form (4.95 Å). This coordination radius can prove that there exist at least some fragments of spiral chain molecule in non-crystalline substances.

As diffraction method alone does not allow to determine atomic structure of non-crystalline matter unambiguously it is necessary to use other methods alongside with diffraction method. Basic details of infrared absorption and Raman scattering spectroscopy of glassy, α-monoclinic and trigonal selenium are given in Table 2.3 (Baratov and Popov, 1990).

The comparison of infrared absorption spectra shows that peaks of absorption in glassy selenium at 95 cm^{-1} and 254 cm^{-1} as well as shoulder at 120 cm^{-1} are in good agreement with bands of fundamental absorption in α-monoclinic form of selenium. This served as the basis for an assumption that a considerable amount of Se_8 molecules is present in glassy selenium. On the other hand deep band of absorption at 135 cm^{-1} in glassy selenium is close to peak 144 cm^{-1} in trigonal selenium and the shoulder at 230 cm^{-1} directly corresponds to the peak in trigonal selenium which is interpreted as proof of presence of spiral chain molecules.

Table 2.3 Features of selenium vibration spectroscopy spectra.

Spectroscopy	Selenium		
	glassy	α-monoclinic	trigonal
Infrared absorption	95 / peak	92–97 / doublet	—
location(cm^{-1}) / type	120 / shoulder	120 / peak	—
	135 / peak	—	144 / peak
	230 / shoulder	—	230 / peak
	254 / peak	254 / peak	—
Raman scattering	110–115 / peak	113 / peak	—
shift (cm^{-1}) / type	140 / shoulder	—	143 / peak
	235 / shoulder	—	237 / peak
	250 / peak	250 / peak	—

The analysis of Raman scattering (Table 2.3) leads to similar conclusions. Glassy selenium spectrum is characterized by a wide peak of complicated form with the maximum at 250 cm^{-1} and a shoulder at 235 cm^{-1}, the position of which covers the peaks at 237 and 250 cm^{-1} on the spectra of trigonal and monoclinic forms. Other peaks of glassy selenium spectrum are close to fundamental modes of crystalline forms as well. Thus, when the results of IR- and RS-spectroscopy are interpreted, the structure of glassy selenium is viewed as a rule as a mixture of molecular forms of monoclinic and trigonal modifications with considerable amount of ring molecules Se_8. But such interpretation is not in agreement with other experiments. So, for example, high viscosity of selenium at the temperatures of glass transition range proves predomination of polymeric molecules rather than monomer ring molecules in the substance.

Lucovsky and Galeener (1980) interpreted the Raman scattering spectra of selenium from the position of the disordered chain model. They showed that in the case of disordered chain selenium molecule (Fig. 2.9(c)) all atoms take part in vibrations with the frequency of 256 cm^{-1} (displacement in the bonding plane) and only part of atoms in the fragments of rings (cis-coupling configuration — Fig. 2.8(a)) contributes to the mode of 113 cm^{-1} (displacement perpendicular to the band plane). This explains the considerably higher intensity of peak 250 cm^{-1}in Raman scattering spectra of glassy selenium.

Differential solubility of glassy selenium in CS_2 or CH_2J_2 is used as an alternative method of estimating the ratio of ring Se_8 and polymer Se_n molecules. As these solvents dissolve monoclinic form well and do not dissolve trigonal form of crystalline selenium at all, it is assumed that ring monomers Se_8 are turned into solution. In the early works using this method an amount of matter turned into solution was up to 20–40 %. But later on it was defined more exactly that solution of glassy selenium in CS_2 depends on the degree of illumination of a sample in the process of dissolving which is probably caused by the structural changes in a substance under the influence of radiation. When radiation has no influence on the dissolving of glassy selenium in the mentioned solvents the amount of substance that turns into solution is as a rule less than 5–10 %. This testifies to low concentration of monomer molecules. Predominance of polymer molecules in glassy selenium is also confirmed by high viscosity of the material.

The results of viscosimetry are often used to obtain information about degree of selenium chain molecule polymerisation. Generally the viscosity of selenium should be determined by the following factors: strength of intermolecular and intramolecular interactions, the ratio of the amounts of ring and chain molecules and the degree of polymerisation of the latter, types and concentration of defects. The level of influence that these factors have on viscosity of selenium will be different under various temperatures. The analysis of temperature dependence of non-crystalline selenium visosity (Popov, 1980) showed that at the range of temperatures 60°–80 °C the visosity of selenium is determined mainly by two factors: intermolecular interaction and molecular structure of material. So in this temperature range the degree of chain molecule polymerisation (the amount of eight-atom monomers in a molecule) may be determined by the expression:

$$\lg\eta = A(T)P^{1/2}, \tag{2.17}$$

where $A(T)$ is factor dependent on the temperature of measuring, P is degree of polymerisation molecules and η is viscosity. As a rule degree of polymerisation value is within 10^3–10^4.

The examined results of research of glassy selenium structure allow to determine accurately enough the short-range order in atomic arrangement, as well as to confirm low content of monomer ring molecules; it also allows to estimate the degree of polymerization of chain molecules and finally to select most probable models of chain molecular structure — the models of free rotating and disordered chains. At the same time there is not enough data to describe spatial disposition of atoms in the material. As mentioned in Sec. 2.4 this task is completed with the help of simulation of non-crystalline material structure.

Vasil'eva *et al.*, (1982) used the gradient method for simulation of structure of glassy selenium layers obtained by vacuum evaporation method at different substrate temperatures. The typical distribution of bond lengths, bond angles and dihedral angles for one of the models are shown in Fig. 2.10. The distribution of the bond lengths in the model has clear maximum that corresponds to both form and disposition of the first maximum of the experimental radial distribution function. The spread of bond length values increases with the decrease of substrate temperature within forming layers. The distribution of bond angles is represented by curves with the range of values between 60° and 180° with the maximum in the interval of 100°–110°. Although the difference between bond angle average values in the model and the value of 105° characteristic for crystalline form increases slightly with the decrease of the substrate temperature, the distribution of bond angles as a whole in models produced at different substrate temperatures differ only slightly. The values of dihedral angles are in the range of 10° to 180° and their distribution does not have explicit maxima, which confirms the presence of free rotation chain model with arbitrary value of dihedral angles in researched samples.

Analysis of the data proves that the structure of investigated layers of glassy selenium consists of disordered chain molecules with small predomination of *trans*-coupling configuration in comparison with *cis*-coupling configuration. The level of structural ordering increases with the increase in the substrate temperature. The latter is proved not only by statistic characteristics of models but also by the analysis of their energy characteristics: total energy of system and its four components (energy of distortion of bond lengths, bond angles, dihedral angles and energy of van der Waals interaction). It is necessary to point out that the main difference in the meaning of total energy is determined by the difference of the van der Waals component, that is, the difference of energy of intermolecular interaction. It corroborates that the changes of degree of structural ordering with the changes of the preparation conditions are determined mainly by the changes in the mutual packing of molecules. That is in medium but not in the short-range order in atomic arrangement.

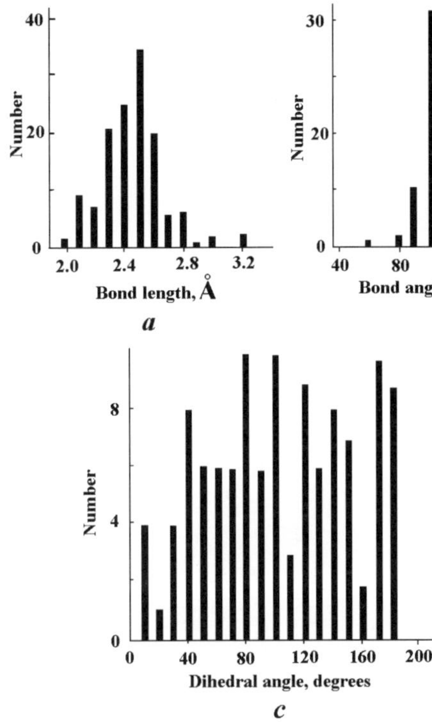

Figure 2.10. Distribution of bond lengths (a), bond angles (b) and dihedral angles (c) in the model of a–Se.

2.5.2 *Atomic Structure of Chalcogenide Glasses*

The structure of non-crystalline arsenic chalcogenides is explained on the basis of continuous network concepts. In terms of chemical order, i.e. correlation between heteropolar (arsenic–chalcogen) and homopolar (arsenic — arsenic, chalcogen — chalcogen) chemical bonds, two extreme cases are possible: completely random network, in which chemical bond distribution is purely statistic, and chemically ordered network where heteropolar bonds realise anywhere if it is permitted by the chemical composition and by the requirement of continuity of the network. In real materials there is some intermediate position determined by the difference of energy value of various bonds, the proportions of components and to a large extent by the conditions of material preparation. For example, according to the data of different researchers concentration of homopolar bonds in films As_2Se_3 differs from 10–15 to 35%.

Most investigations of glassy arsenic chalcogenides structure were carried out on materials of stoichiometric composition As_2X_3 (where X is a chalcogen: sulphur, selenium, tellurium). Particular feature of such compounds is a laminated atomic distribution and the predominance of covalent atomic bonds in layers that is confirmed by isolation of the first peaks of RDF. The basic structural unit is pyramidal blocks $AsX_{3/2}$.

There are fewer investigations devoted to the structure of non-stoichimetric compositions of non-crystalline arsenic chalcogenides. The pyramidal blocks in glasses rich in selenium are assumed to be connected to each other with additional atoms of selenium that increase their free mutual orientation. Free orientation of pyramidal blocks decreases and horizontal length of layers increases in line with growth of arsenic content.

Direct interpretation of area of RDF peaks to determine coordination numbers within binary compositions is possible only for samples of stoichiometric compositions (if we suppose that there are only heteropolar bonds) for the first peak. In order to correctly interpret the results of diffraction investigation of non-stoichiometric binary compounds one needs additional information about interaction of atoms in the substance within an elementary cell due to presence of homopolar bonds in non-stoichiometric compounds. Such information is obtained from spectra of Raman scattering that allows to identify structural units of the investigated substance.

The spectra of Raman scattering (RS) of glassy arsenic selenides are given in Fig. 2.11 (Popov, 2004). Spectrum RS of glassy arsenic consists of peaks at 200, 240, 285 cm^{-1} and a system of peaks in the region of 110–160 cm^{-1}. Well prepared volume specimens of glassy As_2 Se_3 as a rule have only one peak at 227cm^{-1} that is evidence to presence of one kind of structural unit $AsSe_{3/2}$. This in turn confirms correct interpretation of RDF of this compound based on the assumption about existence of heteropolar bonds only.

Comparison of Raman scattering spectra of specimens of system As–Se with different components ratio shows that the peak of stoichiometric compound As_2Se_3 at 227 cm^{-1} is in spectra of all specimens to $As_{65}Se_{35}$. However, intensity of this peak decreases with the increase of arsenic content and only the shoulder is observed in spectra of specimens $As_{65}Se_{35}$ under the given frequency shift. A small excess of arsenic content over stoichiometric composition (sample $As_{55}Se_{55}$ leads to the appearance of a peak at 160 cm^{-1} and a shoulder in the region 240 cm^{-1}. Emergence of these details in the spectrum is linked to the origin of bonds As–As in the glass network. However, absence of a considerable number of peaks characteristic to RS spectrum of non-crystalline arsenic in the spectrum of $As_{45}Se_{55}$ testifies to the absence of structural arsenic units (atom As linked with 3 atoms As) in this sample. Persistence of a large intensity peak at 227 cm^{-1} along with the emergence of the above mentioned details allows to assume that pyramidal blocks $AsSe_{3/2}$ remain predominant structural units in a sample of composition $As_{45}Se_{55}$ and that new structural units appear alongside in which one atom of selenium is replaced by one atom of arsenic — $As(Se_2As)_{1/2}$. Rise in arsenic content to 50% makes a shoulder at 240 cm^{-1} in RS spectrum more expressed and leads to a series of peaks in the zone of 100–160 cm^{-1} that could be a reflection of the fact that the amount of structural units $As(Se_2As)_{1/2}$ and possibility of presence of structural units consisting of only one atom of selenium $As(As_2Se)_{1/2}$ increases. At the same time the absence of a peak at 200 cm^{-1} in RS spectra typical to non-crystalline arsenic testifies to absence of arsenic structural units in this matter. RS spectra of samples with 60 and 65% arsenic are characterized by great reduction of intensity of the peak at 227 cm^{-1} typical to As_2Se_3 (that confirms a small concentration of structural units $AsSe_{3/2}$), rise of intensity of peaks at 240 cm^{-1} and at a range of 110–160 cm^{-1}. Besides, peaks at 200 and 285 cm^{-1} become apparent and thus all peaks typical to non-crystalline arsenic are present in these spectra. The obtained data testifies to predominance of structural units $As(Se_2As)_{1/2}$ and As $(As_2Se)_{1/2}$ in these substances as well as to existence of arsenic structural units in them.

Analysis of Raman scattering results in substances of systems As–S and Sb–S at a range of compositions from $Sb_{36}S_{64}$ to $Sb_{43}S_{57}$ shows that changes of spectra together with change of components ratio are analogous to the ones considered for system As–Se.

Additional information about the structure of arsenic chalcogenide glasses can be obtained by indirect methods by measuring various physical and chemical properties of the matter. Qualitative results of alteration of heteropolar and homopolar bonds ratio in the materials of systems As–Se, As–S can be obtained using the method of differential solution. It is known that alkaline solvents have more impact on heteropolar bonds (As–Se, As–S) than on homopolar ones (As–As, Se–Se, S–S) in the process of dissolving of arsenic sulphides and selenides.

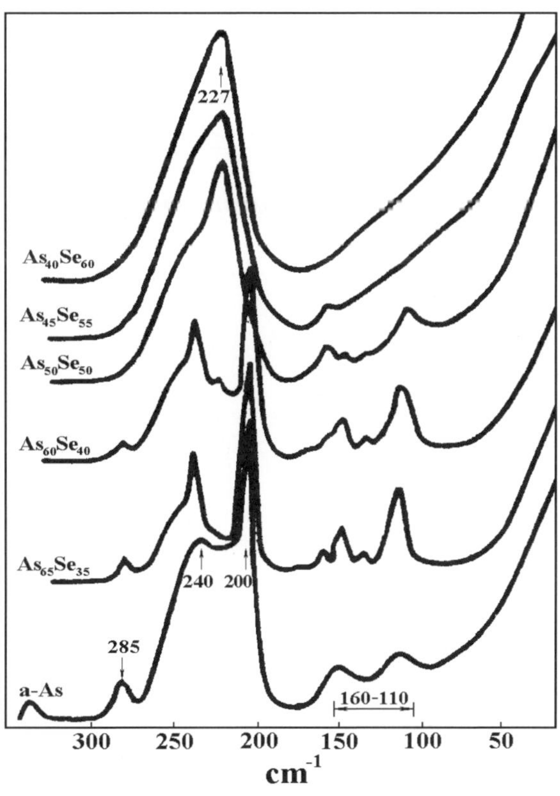

Figure 2.11. Raman spectra of glassy arsenic chalcogenides (a–As after Graves *et al.*, 1979).

Investigation of As–Se materials solubility in 10% KOH showed that speed of dissolving falls significantly and the quantity of insoluble residue rises with the increase of arsenic content over stoichiometric composition ratio. It conforms to the conclusion made on the basis of analysis of Raman scattering spectra that the amount of As–As bonds increases in this case.

The chalcogenides of the IV group elements have smaller regions of glass formation and demand harder conditions of melt quenching in comparison with arsenic chalcogenides. In their glassy state, silicon and germanium have valency 4. The main features of structural changes with different composition are similar to those for arsenic chalcogenides observed above. Tetrahedrons $GeX_{4/2}$, $SiX_{4/2}$ that consist of a central atom germanium (silicon) bonding with 4 atoms of chalcogens are the structural units of composition GeX_2, SiX_2. In compositions enriched by chalcogens a number of tetrahedrons is proportional to atomic concentration of the IV group elements and surplus atoms of chalcogen are united in chains linking tetrahedrons among themselves. Under higher concentration of chalcogen separate molecules of the latter appear and form a solid solution with structural network of material.

2.5.3 Atomic Structure of Amorphous Silicon

Tetrahedral bonded semiconductors (Si, Ge, SiC and others) in non-crystalline state may be prepared in the form of thin films only. The first coordination number in these non-crystalline films equal 4 ± 0.05 and first coordination sphere radius corresponds with bond length in crystalline form (with deviation of no more than $\pm$ 0.06 Å). The same applies for the mean value of bond angle. This indicates that tetrahedral structure and short range order in non-crystalline states remains the same as in crystals. However, mutual packing of tetrahedrons differs in the case of non-crystalline and crystalline states. It leads to the loss of long range order in non-crystalline state and is an essentially different form of radial distribution functions beginning on the third maximum. Taking these facts into consideration a number of structural models have been created (Secs. 2.2 and 2.4).

It is necessary to note that all these models represent idealized atomic structure of the material because neither dangling bonds inside model nor inside clusters were permitted. In reality, rigid covalent network of tetrahedral bonded disordered semiconductors contains a considerable amount of dangling bonds that strongly influence properties of the material. Addition of hydrogen atoms in amorphous silicon films leads to a sharp decrease of dangling bond concentration and to changes of material's properties (see Chapter 3). Content of hydrogen in amorphous silicon films may amount to 30 at.%. In fact it is a new material: silicon–hydrogen alloy or hydrogenated amorphous silicon (a–Si:H). In comparing a–Si:H structure to a–Si structure it is necessary to answer to following questions. (1) What is the influence of hydrogen atoms on silicon's structural network? (2) What are hydrogen atom surroundings in the structural network of hydrogenated amorphous silicon? Hydrogen atomic scattering factor is negligible in comparison with

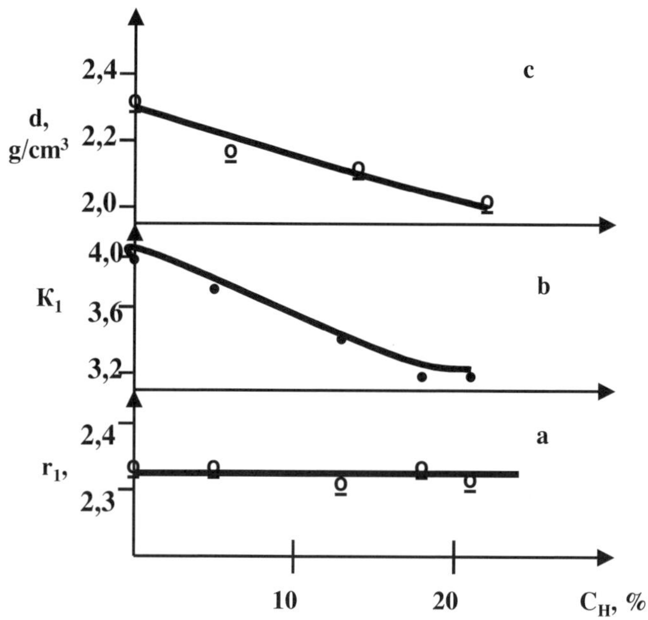

Figure 2.12. Dependencies of the first coordination sphere radius (a), the first coordination number (b) and the density (c) of hydrogenated amorphous silicon on hydrogen concentration.

scattering factor of silicon atoms in the case of diffraction experiment. Therefore hydrogen atoms do not give visible contribution in diffraction pattern. That is why diffraction methods make it possible to calculate radial distribution function of silicon atoms in a–Si:H. Vibration spectroscopy methods are usually used to answer to second question.

The dependencies of the first coordination number, the first coordination sphere radius and the density of hydrogenated amorphous silicon on hydrogen concentration are shown in Fig. 2.12. (Aivasov *et al.*, 1995). One can see that the first coordination sphere radius remains constant with increasing hydrogen concentration. At the same time values of the first coordination number and density of material decrease.

The reduction in the material's density as a result of hydrogen concentration increase is obvious. The decrease in the first coordination number is conditioned by formation of bonds Si–H instead of some of the bonds Si–Si. The following configurations of atoms are possible in this case: silicon atom bonded with three atoms of silicon and one atom of hydrogen (Si–H), silicon atom bonded with two atoms of silicon and two atoms of hydrogen (Si–H_2) and silicon atom bonded with one atom of silicon and three atoms of hydrogen (Si–H_3). These configurations have different group frequencies (vibration modes) in the infrared absorption and Raman scattering spectrums. Thus location and intensity of appropriate peaks in vibration spectrums make it possible to identify and estimate concentration of the structural units mentioned above. For example, hydrogen atom concentration in various structural units may be determined from infrared absorption data as follows:

$$C_H = \frac{1}{\Gamma}\int \frac{\alpha(v)}{v}dv \tag{2.18}$$

where C_H is hydrogen atom concentration, ν is wave number, α is absorption coefficient and Γ is trapping section.

The ratio of various structural units is determined primarily by methods and conditions of material preparation and exerts determinant influence on the material's properties.

Chapter Three

Electronic Structure and Properties of Disordered Semiconductors

3.1 ELECTRONIC STRUCTURE

3.1.1 Localized States in Disordered Semiconductors

Solid state physics provides us with a simple understanding of the electronic structure of crystalline semiconductors. However, solid state physics is based on periodic crystal lattice (see Chapter 1). This means it has nothing to say about electronic structure of non-crystalline semiconductors that do not have the periodic lattice. On the other hand the main features of electronic structure of atoms in the solid are determined by interaction between nearest neighbours, or in other words, by short range order of atomic arrangement. One could deduce that if the short range order remains the same in crystalline and non-crystalline states of material then main features of electronic structure of the material remain similar as well. Of course, similar does not mean the same. Absence of long range order leaves traces on the spectrum of electronic states of non-crystalline semiconductors. The main distinguishing feature of electronic spectrum of non-crystalline semiconductors is localized states.

There are two definitions of localized states suggested by P. Anderson and by N. Mott. The first one is: the state is localized if electron with energy $E \pm \Delta$ located in volume large enough to satisfy the uncertainty principle, does not diffuse from this volume. N. Mott has mentioned the other feature of localized states: the contribution of electrons in localized states to direct current conductivity reduces to zero at low temperatures.

From the point of view of physics this can be explained as follows. The density of electronic states at the conduction and valence bands is high. It means that the states at the same energy are separated by very thin barriers. Electrons can tunnel through these barriers under the influence of concentration gradient or weak electrical field (Fig. 3.1(a)). If density of states decreases the states of the same energy become well separated in the lattice site, so that tunneling probability between these sites is zero (Fig. 3.1(b)). The states become localized. Electrons in these states cannot diffuse (Anderson definition) or take part in electrical

Disordered Semiconductors: Physics and Applications by Anatoly Popov

www.panstanford.com
978-981-4241-76-2

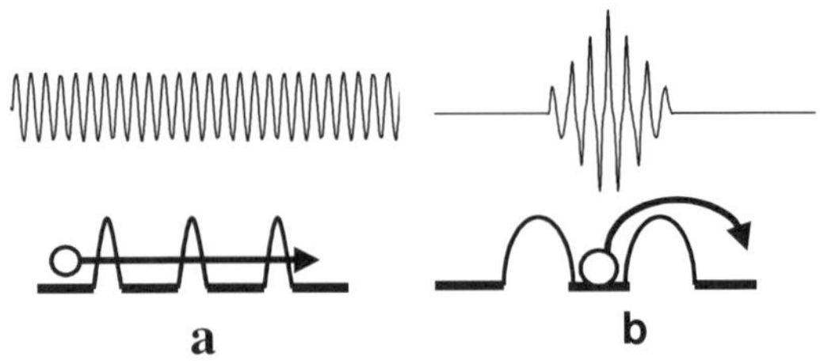

Figure 3.1. The wave functions and transitions of electrons in the delocalized (extended) states (a) and localized states (b).

conductivity (Mott definition). Only over barrier electron transitions are possible in this case.

3.1.2 Models of Energy Bands

Localized states mentioned above are present not only in disordered semiconductors but also in crystals as well. Donor and acceptor states in crystals are localized states. For example, an electron at donor level cannot transfer in crystal until its excitation to conduction band. Another example of localized states in crystals is deep levels in the gap caused by crystalline lattice defects. Presence of a point defect (vacancy or interstitial defect) leads to the change in neighboring atoms' interaction and to the shift of atomic orbitals' energy.

The absence of long range order in non-crystalline semiconductors leads to shifts of electronic states in conduction and valence bands as well. Some of these states may shift to the gap. As a result the tails of conduction and valence bands in the gap are formed (Fig. 3.2) (Gubanov, 1963). The tail states are localized and are separated at E_c and E_v from the extended (delocalized) states.

This model is rather clear from the physics point of view but it cannot explain some experimental facts. For example, absence of localized states in the middle of the gap does not explain the effect of so-called fixation of Fermi level in the middle of the gap which is experimentally observed in chalcogenide glasses.

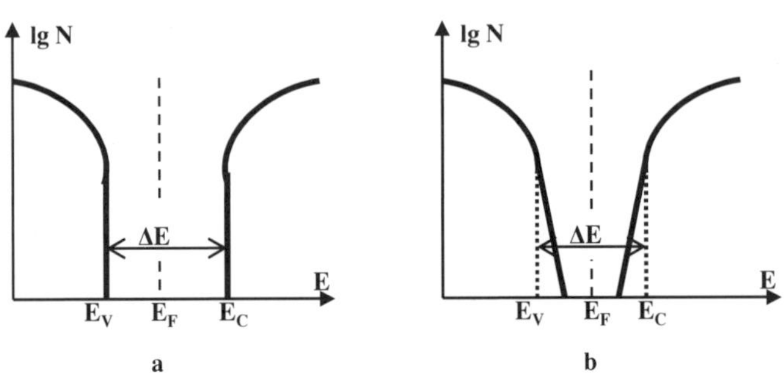

Figure 3.2. Schematic illustration of energy bands in the case of crystaline (a) and non-crystaline (Gubanov model) (b) semiconductors.

It is known that Fermi level position depends on temperature. In turn the value of Fermi level temperature shift depends on density of electronic states. The shift has bigger value in the case of lower density of electronic states in the region of Fermi level position. However Fermi level in non-crystalline semiconductors is situated near the middle of gap and its position only changes slightly even with significant temperature variation. This can only happen when a small shift of Fermi level position leads to a considerable change of ratio between occupied (lower E_F) and empty (higher E_F) states or when the density of states in the region of Fermi level position is high. But density of electronic states in the middle of gap equal zero in Gubanov model (Fig. 3.2(b)).

That is why Cohen, Fritzscher and Ovshinsky (CFO) (1969) proposed another model of energy bands for multicomponent chalcogenide glasses. They suggested that due to additional chemical disorder of these materials, tails of localized states have larger length and overlap in the middle of forbidden gap (Fig. 3.3(a)). But in this case the concept of forbidden gap makes no sense at all because electronic states are present at any energy. That is why the authors introduced forbidden gap of mobility instead of forbidden gap of electronic states (Fig. 3.3(b)). Forbidden gap of mobility is limited to values of energy E_c and E_v,which separate extended (high mobility) and localized (low mobility) states. Mobility in localized states tends to zero at low temperatures.

Overlapping of valence and conduction band tails leads to formation of charged centers (positively charged higher Fermi level position and negatively charged lower Fermi level position). It explains the effect of fixation of Fermi level in the middle of the gap. But this model is in contradiction with some other experimental data because density of states in forbidden gap is too large (for example some optical properties, temperature dependence of AC conductivity and others).

The next step was made by Davis and Mott (1970). Their model (Fig. 3.4) has short tails of localized states and narrow region with high density of localized states in the middle of gap, which fixes Fermi level. This region cannot be

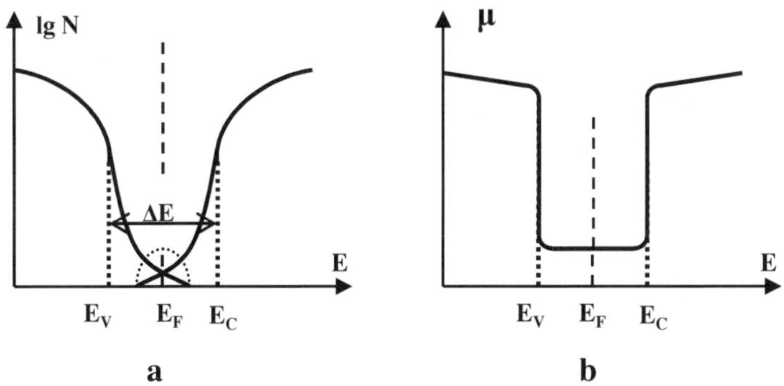

Figure 3.3. Schematic illustration of density of states (a) and carrier mobility (b) in CFO model.

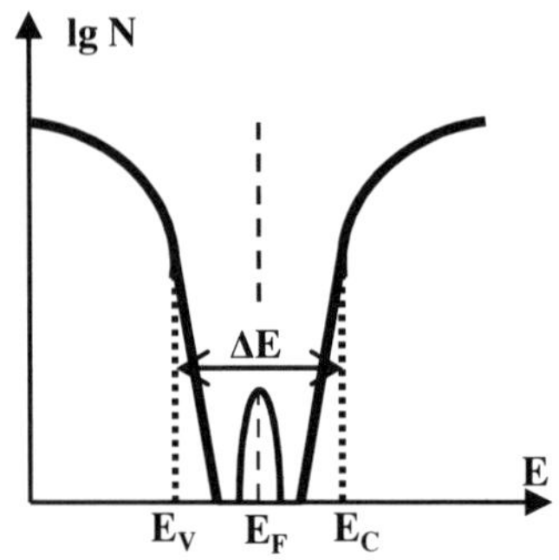

Figure 3.4. Schematic illustration of density of states after Mott-Davis model.

explained by fluctuation shift of electronic states from conduction and valence bands. The authors suggested that it consists of electronic states of structural defects. This model is generally used now. Let us consider possible types of intrinsic defects in non-crystalline semiconductors.

3.1.3 Defect States in Disordered Semiconductors

The absence of long range order causes much more variety of defect types in non-crystalline solids in comparison with crystals. Point defects in crystals (vacancy or interstitial defects) represent the variation of coordination number of several atoms simultaneously. For example vacancy in tetrahedral crystalline lattice leads to formation of four atoms with coordination number of three (Fig. 3.5(a)). By contrast, the variation of coordination number of three, two or one atoms is possible in non-crystalline tetrahedral material (Fig. 3.5(b)).

Equilibrium concentration of defects both in crystals and non-crystalline materials is determined by free energy minimum condition:

$$G = H - TS \tag{3.1}$$

where H is enthalpy and S is entropy. Defect formation leads to increase in both enthalpy (on the value of defect formation work) and entropy because entropy

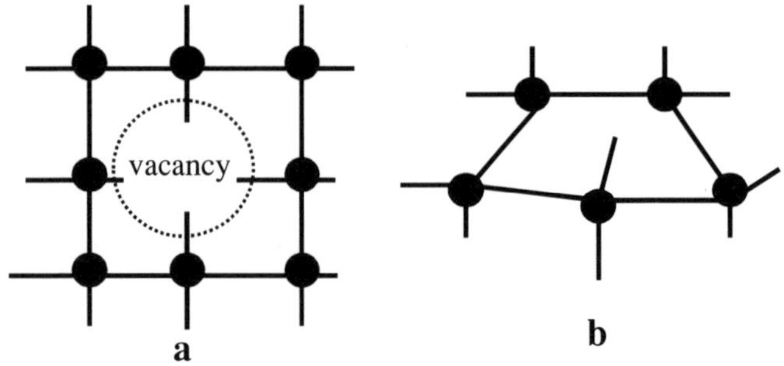

Figure 3.5. Variation of atom coordination number in point defect region in the case of tetrahedranl crystal (a) and tetrahedral non-crystalline material (b).

may be written as statistical weight of system:

$$S = k \cdot \ln \frac{N!}{(N-n)!} \tag{3.2}$$

where n is concentration of defects, N is number of possible places of defect distribution and k is Boltzmann constant. Therefore equilibrium defect concentration differs from zero at any temperatures except 0 K and increase exponentially with rising temperature. In the case of disordered solids concentration of defects has to be more than equilibrium concentration because of nonequilibrium preparation conditions. Experimental results of drift carrier mobility, photoconductivity, value of Stokes shift, effect of Fermi level fixation and others confirm the presence of high defect concentration and appropriate localized states in the gap of disordered semiconductors.

Let us consider electronic structure of the common type of defects in non-crystalline semiconductors — abnormal coordinated atom. The example of this type of defect is the atom with dangling bond at the end of chalcogen chain molecule. The atom inside of chain molecule forms two covalent bonds. As a result two p-electrons pass on bonding orbitals and two p-electrons remain on non-bonding orbitals (Fig. 1.7(b)). The atom at the end of chain molecule forms only one covalent bond. Therefore only one p-electron passes on bonding orbital and three p-electrons remains on non-bonding orbitals (Fig. 3.6). Because there is an odd number of electrons on non-bonding orbitals this defect has to have non-paired spin. Taking into consideration high concentration of defects and non-paired spins it is reasonable to expect the presence of electron spin resonance (ESR) signal in these materials. However ESR signal is absent in chalcogenide glasses and glassy selenium in normal conditions.

The absence of ESR signal means that localized states in the gap of chalcogenide glasses are either empty or occupied twice. But it means that these centers have a positive (non-paired electron removed) or negative (extra electron added) electric charge. It is rather difficult to explain such situation from the point of view of crystal physics. Charge state in crystals has greater energy in comparison with neutral state because of repulsive Coulomb interaction. The energy difference between charge and neutral states is named correlation energy U and always has positive value in crystals (Fig. 3.7).

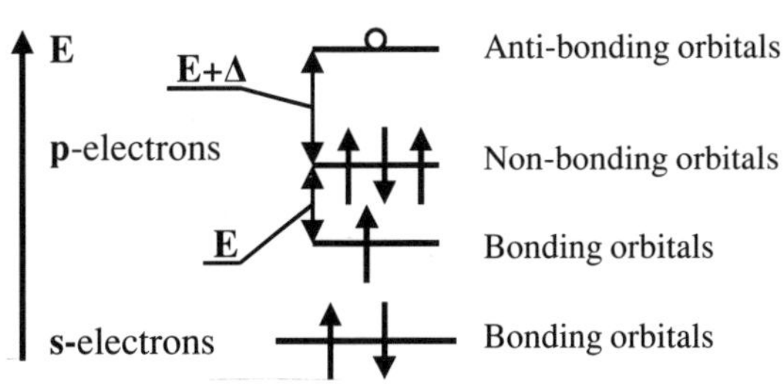

Figure 3.6. Schematic diagram of orbitals in the case of atom at the end of chalcogen chain molecule.

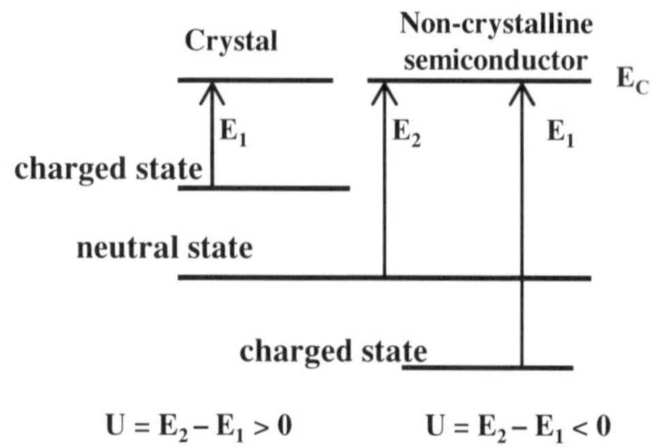

Figure 3.7. Schematic illustration of correlation energy determination(see text).

Anderson (1975) has suggested the concept of negative correlation energy. He pointed out that in the case of flexible structural network of chalcogenide glasses electron-phonon interaction gives rise to an attractive interaction between electrons overcoming the repulsive Coulomb interaction. As a result charged defects have to have less energy as compared with neutral defects and correlation energy becomes negative.

Concerning the nature of center with negative correlation energy Street and Mott (1975) proposed that dangling bonds are negative correlation energy defects. Neutral dangling bonds (D^0) become positively (D^+) and negatively (D^-) charged as a result of network distortion, which brings about negative correlation energy. That is why

$$2D^0 \rightarrow D^+ + D^- \tag{3.3}$$

is an exothermal reaction. In other words the first ionization of negatively charged center ($D^- \rightarrow D^0 + e$) requires energy E_1 greater than energy of second ionization E_2 ($D^0 \rightarrow D^+ + e$) (Fig. 3.7). As a result correlation energy $U = E_2 - E_1$ has negative sign and defects in chalcogenide glasses have to be positively or negatively charged. Neutral state D^0 can be achieved only by excitation.

Electron transitions between valence band and defect states are shown in Fig. 3.8(a). The lower parabola represents D^+center and filled valence band. The upper parabola represents D^0 center with a hole at the top of the valence band. Transition ***A*** corresponds to excitation of electron from the valence band into D^+ center, leaving a hole in the valence band. After the excitation of electron into D^+, the charge of state is changed ($D^+ + e = D^0$). This leads to rearrangement of structural network and to the decrease in energy of the polaron value w^+. Transition ***C*** corresponds to recombination of D^0 with a hole in the valence band and the appropriate rearrangement of structural network and energy decrease of w^+. Both transitions are vertical (optically allowed) transitions, following the Frank-Condon principle. Thus Stokes shift is equal to doubled polaron energy ($E_A - E_C = 2w^+$). Transition ***B*** corresponds to thermal excitation of electron in the valence band intoD^+ center.

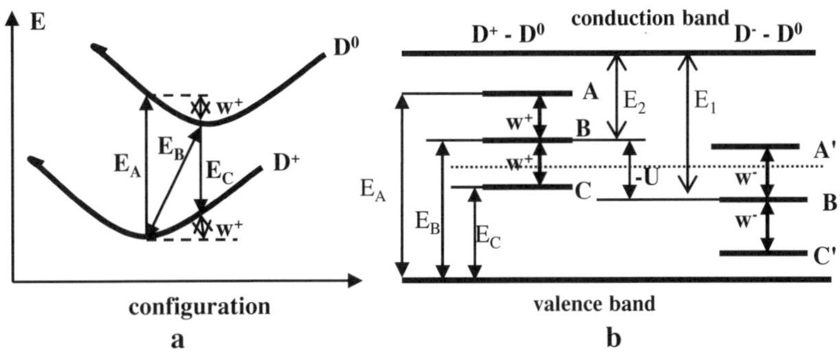

Figure 3.8. Configuration (a) and level (b) diagrams for defect levels with negative correlation energy.

Transition energies E_A, E_B and E_C are shown in level diagram (Fig. 3.8(b)). Similar levels for the case D^-–D^0 transitions are shown on the right side of figure. The energy difference between ***B*** and ***B′*** levels $(E_2 - E_1)$ corresponds to the negative correlation energy.

Kastner, Adler and Fritzscher (1976) have analyzed defects D^+ and D^- from the point of view of chemical bonding theory. They proposed that positively charged defect D^+ has minimum energy in the case of three-fold coordinated (over-coordinated) chalcogen atom C_3^+. On the contrary, negatively charged defect D^- has minimum energy in the case of one-fold coordinated (under-coordinated) chalcogen atom C_1^-. (Superscripts and subscripts denote charge states and coordination number respectively). In this case Eqn. (3.3) may be rewritten as follows:

$$2C_2^0 \rightarrow C_3^+ + C_1^- \tag{3.4}$$

It is necessary to note that the total number of atom covalent bonds in reaction (3.4) does not change (and equals four). However the coordination number of each atom changes. That is why this type of defects was called **valence alternation pairs** (VAP). Let us consider the VAP formation process in glassy selenium.

Two fragments of selenium chain molecules and electron configuration of selenium atoms are shown in Fig. 3.9. As mentioned above, atoms inside the chain have two $p-$electrons on bonding orbitals and two $p-$electrons on non-bonding orbitals. Atoms at the end of the chain have only one $p-$electron on bonding orbital and three $p-$electrons on non-bonding orbitals (including one electron with unpaired spin) (Fig. 3.9(a)). The electron with unpaired spin can interact and form covalent bond with lone pair electron of neighboring atom (Fig. 3.9(b)):

$$C_1^0 + C_2^0 \rightarrow C_3^0 + C_2^0 \tag{3.5}$$

as a result three-fold coordinated atom C_3^0 arises. This atom has three electrons on bonding orbitals and one electron on anti-bonding orbital due to the transition of two electrons from non-bonding to bonding orbitals (energy reduction on the value $2E$) and one electron from non-bonding to anti-bonding orbital (energy increase on the value $E + \Delta$). Total energy change is equal $-E + \Delta$. Thus under

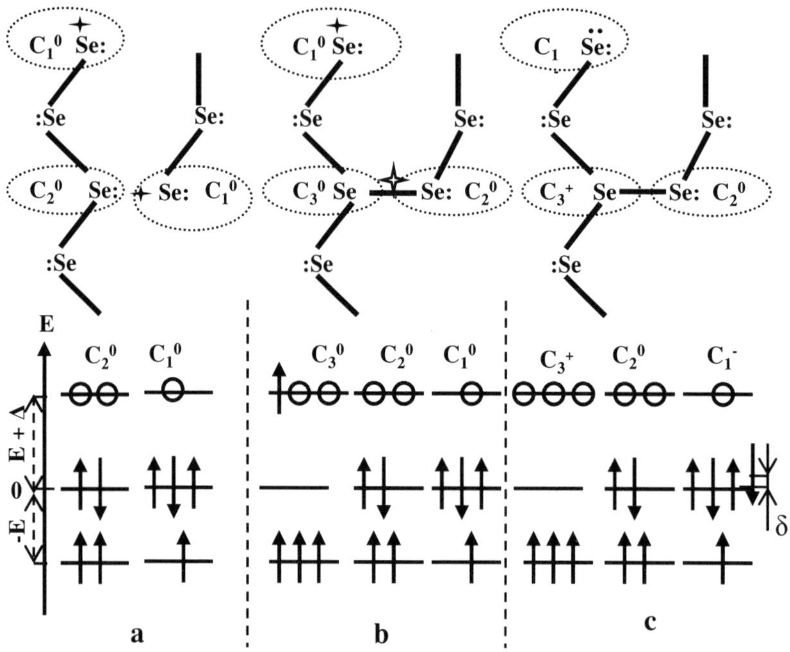

Figure 3.9. Formation of valence-alternation pair in glassy selenium. See the text for elucidation (:lone-pair of electrons, + non-paired electron on non-bonding orbital and ✧ electron on anti-bonding orbital).

the assumption of $\Delta = E$ reaction (3.5) is accompanied by a considerable energy gain.

The energy of an electron in anti-bonding state is close to conduction band. So there is probability of electron transfer from anti-bonding state of C_3^0 atom to non-bonding state of C_1^0 atom (Fig. 3.9 c):

$$\mathbf{C}_3^0 + \mathbf{C}_1^0 \rightarrow \mathbf{C}_3^+ + \mathbf{C}_1^- \tag{3.6}$$

The formation of C_1^- requires additional energy δ associated with Coulomb interaction of electrons on non-bonding orbitals. Thus, if $E > \Delta + \delta$ the formation of valence alternation pairs C_3^+ and C_1^- is an exothermic process and these defects are stable configurations.

Presence of dangling bonds is not an indispensable condition for valence alternation pair formation. The example of **close** $C_3^+ - C_1^-$ pair formation by exciton self-trapping is shown in Fig. 3.10 (Street, 1977). Rearrangement of only one bond and associated distortions of the atoms is necessary for this transformation. The Coulomb interaction makes the energy of close $C_3^+ - C_1^-$ pair even smaller than energy of distant $C_3^+ - C_1^-$ pair. Close $C_3^+ - C_1^-$ pairs are meta-stable states and may help to understand the nature of photo-structural change effect in chalcogenide glasses (see Sec. 6.2.1).

More stable configuration of close VAP states named **intimate valence alternation pairs** (IVAP) has been suggested by Kastner (1977), Kastner and Hudgens

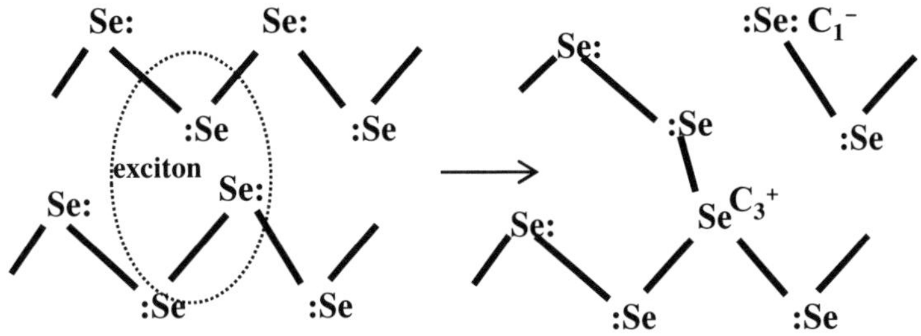

Figure 3.10. Schematic diagram of the transformation of an exciton into close $C_3^+ - C_1^-$ pair.

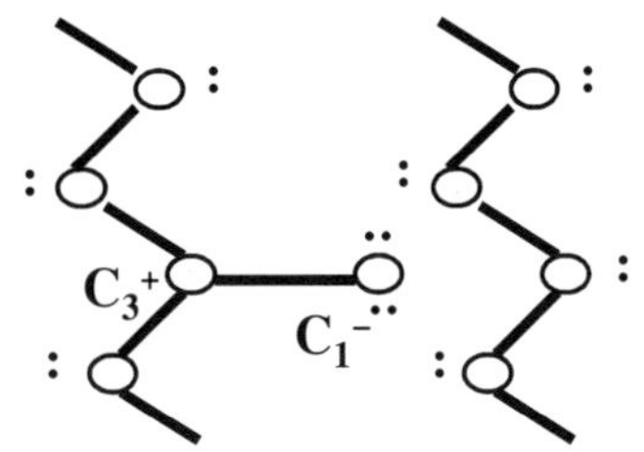

Figure 3.11. Intimate valence alternation pair.

(1978). $C_3^+ - C_1^-$ pair forms diamagnetic complex with the total charge equaling zero in this model (Fig. 3.11).

An alternative model of defect states in chalcogenide glasses is the model of **quasi-molecular defects** suggested by Popov N. (1980, 1981). The model is based on the electron-surplus bond concept (Pimentel and Spratley, 1970) and assumes presence of multi-centre orbital-deficiency bonds in chalcogenide glassy semiconductors. Formation of simplest three-center quasi-molecular defect in selenium is shown in Fig. 3.12.

Two electrons from non-bonding orbitals of atom number 1 (points in the circle in the figure) and two electrons from bonding orbitals of atoms 2 and 3 (solid line in the figure) form orbital-deficiency bond of three selenium atoms (dotted line in

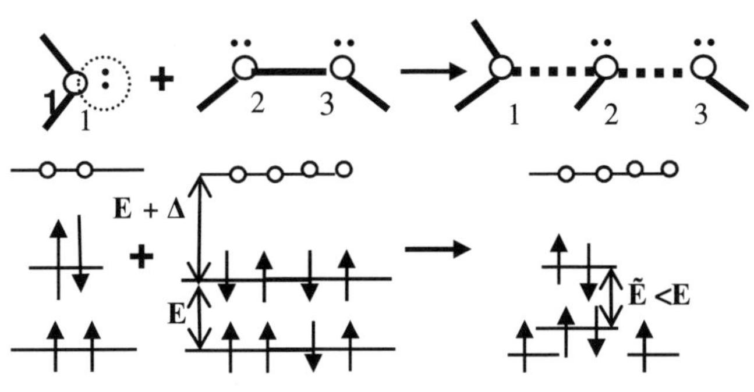

Figure 3.12. Schematic diagram of the quasi-molecular defect formation.

the figure). At the same time these electrons occupy two non-bonding orbitals and two bonding orbitals (energy $\tilde{E} < E$) as before bond formation. Thus two atoms with coordination number three and one atom with coordination number two appear instead of three two-fold coordinated atoms. Quasi-molecular defect has no electrical charge but has dipole moment because of redistribution of electron density. The constant value of electrons on bonding and non-bonding orbitals as well as absence of electrons on anti-bonding orbitals leads in the author's opinion to less defect formation energy in comparison to VAP model.

The VAP, IVAP and quasi-molecular defect models do not negate but supplement one other. It is not unreasonable to suppose that all types of defect and transformation of one type of defect into another are possible in chalcogenide glassy semiconductors.

Creation of defects like valence alternation pairs is possible in the atoms of fifth group of periodic table (arsenic, antimony) as well. But in this case, the transformation is more complicated because over-coordinated defects can only arise in the case of orbital hybridization.

The normal state of V group atom (pnictide) is ${}_pP_3^0$, where p-electrons form three covalent bonds (Fig. 3.13(a)). It has been shown (Elliott and Davis, 1979) that the lowest energy defects are under-coordinated defect ${}_pP_2^-$ (Fig. 3.13(b)) and over-coordinated defect ${}_{sp}P_4^+$. Over-coordinated defect forms due to *sp*-hybridization (Fig. 3.13(c)).

In the case of chalcogenide glasses that contain atoms of both group elements, all defects (C_3^+, C_1^-, P_4^+, P_2^-) may be present. But the concentration of different types of defects depends on defect formation energy. It has been shown that the main under-coordinated defect is C_1^-. Over-coordinated defects C_3^+ and P_4^+ have approximately equal values of formation energy.

The presence of lone-pair electrons on non-bonding orbitals is an indispensable condition of valence alternation pairs formation. This is why this type of defects is possible in the case of V and VI group elements of periodic system but it cannot be formed on atoms of IV group in the absence of lone-pair electrons. Consequently only under-coordinated defects may be present in this case.

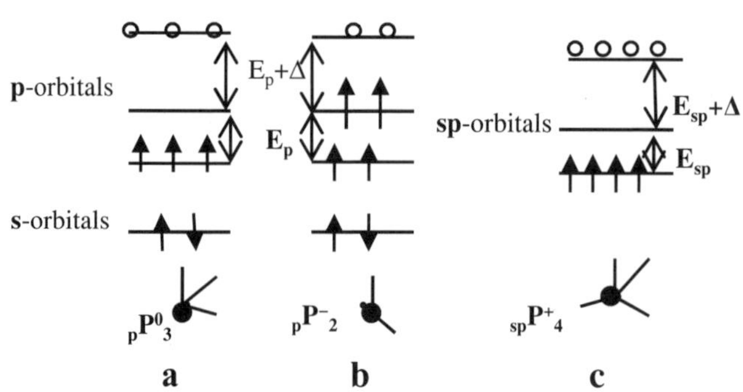

Figure 3.13. Energy levels of pnictide atom in normal (a) under (b) and over (c) coordinations.

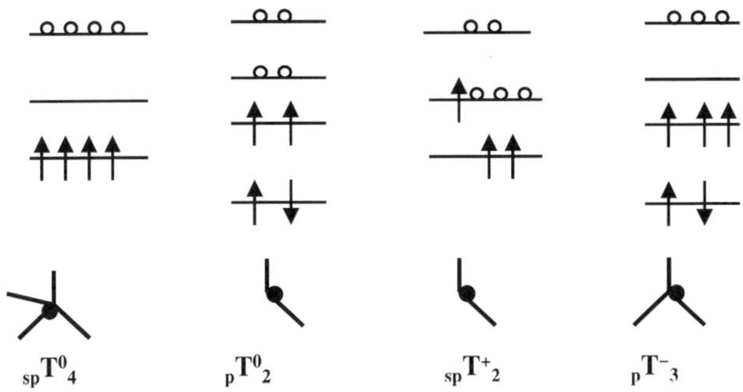

Figure 3.14. Energy levels of silicon atom in normal (a) and defect (b, c and d) configurations.

Electronic structure of feasible defects in amorphous silicon has been analyzed by Adler (1978). Silicon atom in normal configuration has coordination equaling four as a result of sp^3- hybridization of electron orbitals $_{sr}T_4^0$ (Fig. 3.14(a)). In author's estimate the lowest energy defect in amorphous silicon is the neutral two-fold coordinated atom $_pT_2^0$ (Fig. 3.14(b)). The second level of energy has charged defects $_{sp}T_2^+$ and $_pT_3^-$ (Fig. 3.14(c, d)). $_{sp}T_2^+$ center contains an unpaired spin which gives ESR signal. It is important to note that neutrality of the lowest energy defect signifies the positive correlation energy of its formation. It may be explained by significantly more rigid structural network of amorphous silicon in comparison with chalcogenide glassy semiconductors.

3.1.4 Electronic Structure of Hydrogenated Amorphous Silicon

The distinctive feature of disordered semiconductors is the weak influence of impurity on electrical properties of these materials (see Chapter 4 for details). In the case of amorphous silicon this fact is explained by high density of intrinsic defects due to the presence of a large number of dangling bonds in four-fold coordinated rigid disordered structural network. As a result density of localized states in the gap of amorphous silicon reaches 10^{20} cm^{-3}eV^{-1}. The concentrations of donor or acceptor levels after doping proved to be less than intrinsic defect concentration (Fig. 3.15(a)) and impurities do not influence the Fermi level position and the electrical properties of semiconductor.

Hydrogen introduction in amorphous silicon leads to formation of silicon–hydrogen (Si – H) bonds instead of dangling silicon bonds. Bond creation is accompanied by electron transition from non-bonding orbitals of dangling silicon bond and hydrogen atom to bonding orbitals. Because energy of Si – H bond exceeds the energy of Si – Si bond these bonding orbitals shift to valence band and anti-bonding orbitals are disposed in conduction band (Fig. 3.15(b)). It leads to a sharp decrease of localized state density in the gap of hydrogenated amorphous silicon (Fig. 3.15(a)) and to possibility of impurity level density exceeding that of intrinsic defect levels.

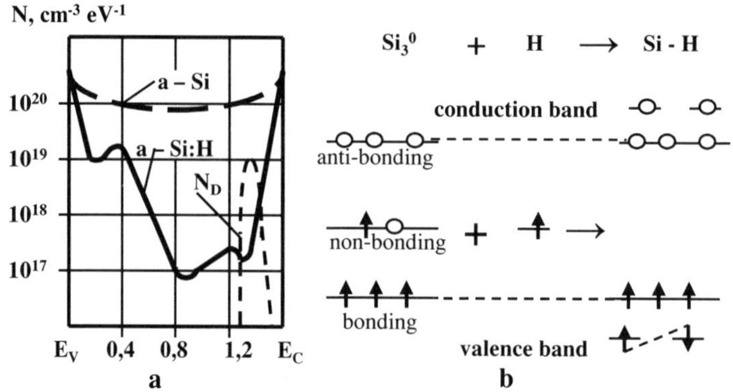

Figure 3.15. Density of localized states in the gap of a–Si and a–Si:H (a) and illustration of Si-H bond formation

3.2 ELECTRICAL PROPERTIES OF DISORDERED SEMICONDUCTORS

Three physical characteristics — electrical conductivity, thermoelectric power and Hall effect are usually used to investigate electrical charge transport in semiconductors. Knowledge of these characteristics and their temperature, frequency and field dependencies allows to determine practically all parameters of electrical charge transport in crystalline semiconductors: type of conductivity, charge carrier concentration, carrier mobility, activation energy of conductivity and others.

Absence of long-range order of atom arrangement in non-crystalline semiconductors exerts strong influence on electrical transport. This is why different theoretical and experimental approaches to electrical transport analyses have been created. There is a number of electrical transport models in non-crystalline semiconductors at present time but only the most common models are considered in this section.

3.2.1 *Electrical Conductivity*

We will discuss the electrical conductivity on the basis of Mott-Davis model of energy bands (see Sec. 3.1.2). Characteristic features of the model are (Fig. 3.16):

- presence of mobility edges at the energies E_C and E_V separating localized and extended states;
- presence of localized state tails of conduction band (between energies E_C and E_A) and valence band (between energies E_V and E_B);
- presence of narrow band of localized states in the middle of gap (between energies E_1 and E_2.

In accordance with this model three types of electron transport are possible:

- carrier transport in extended (delocalized) states (carrier energy more than E_C and less than E_V;

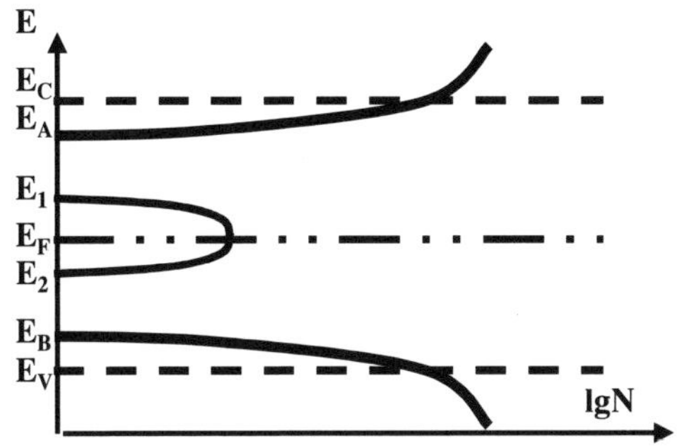

Figure 3.16. Mott-Davis model of energy bands.

- hopping conduction in the tails of conduction band and valence band (carrier energy between E_C and E_A and between E_V and E_B);
- hopping conduction in localized states near the Fermi level (carrier energy between E_1 and E_2).

It is clear that the contribution of each type of transport to material conductivity depends on temperature, frequency of electrical field and so on. Let us begin from temperature dependence of direct current conductivity.

Conductivity in extended states is observed at a sufficiently high temperature because carriers have to excite to delocalized states. In this case conductivity is similar to conductivity of crystalline semiconductors. Temperature dependence of p-type conductivity is given by

$$\sigma = \sigma_0 \exp[-(E_F - E_V)/kT] \tag{3.7}$$

and in the case of n-type conductivity is given by

$$\sigma = \sigma_0 \exp[-(E_C - E_F)/kT] \tag{3.8}$$

where pre-exponential factor $\sigma_0 = 10\text{–}10^3 \Omega^{-1}\ \text{cm}^{-1}$ in the case of non-crystalline semiconductors. Because the value of gap has linear dependence on temperature the temperature dependence of $(E_F - E_V)$ is given by

$$E_F - E_V = (E_F - E_V)_{T=0K} - \gamma T \tag{3.9}$$

where $\gamma = dE/dT$ is band temperature coefficient. So Eqn. (3.7) may be written as follows:

$$\sigma = \sigma_0 \exp(\gamma/k) \exp[-(E_F - E_V)_{T=0K}/kT] \tag{3.10}$$

and plot $\ln\sigma(1/T)$ is the straight line with slope equal $(E_F - E_V)_{T=0K}/k$ and with point of ordinate axis intersection at $\sigma_0 \exp(\gamma/k)$. Carrier mobility on the analogy of crystalline semiconductors slightly decreases with the increase in temperature

$$\mu \sim 1/kT \tag{3.11}$$

and lies in the range from 1 to 10 $\text{cm}^2/(\text{V} \cdot \text{s})$ at 300 K.

Conductivity in the tails of bands is realized by carriers excited to the states in band tails. Because these states are localized electrical transport may be realized only by electron hopping from site to site (Fig. 3.1(b)) with the assistance of phonons (so called hopping conductivity). In this situation mobility of carriers is much smaller in comparison to conductivity in extended states (approximately 10^{-2} cm^2/(V·s) at 300 K) and depends on the value of activation energy of the hop between two sites:

$$\mu \sim (1/kT)\exp(-W_1/kT) \tag{3.12}$$

where w_1 is activation energy of hops. Thus there is exponential dependence of mobility on temperature in this case. As a result, activation energy of conductivity is changed as well. Conductivity of p-type is given by

$$\sigma = \sigma_1 \exp[-(E_F - E_V + w_1)/kT] \tag{3.13}$$

and n-type conductivity is given by

$$\sigma = \sigma_1 \exp[-(E_A - E_F + w_1)/kT] \tag{3.14}$$

where pre-exponential factor is $\sigma_1 = (10^{-2} - 10^{-4})\sigma_0$ due to low mobility value. So in this case activation energy of conductivity is the sum of activation energy of carrier concentration $(E_F - E_V)$ and activation energy of mobility (w_1).

Conductivity in localized states at the Fermi level is a hopping conductivity as well. There are two types of carrier transport in this case: nearest — neighbor hopping and variable — range hopping.

If temperature is high enough, then hopping is most probable between nearest neighbors with activation energy w_2 (Fig. 3.17) because the transfer integral between these sites is rather large. This hopping is called nearest — neighbor hopping. This type of hopping is similar to the conduction through impurity levels in heavily doped compensated crystalline semiconductors. Temperature dependence of conductivity due to nearest-neighbor hopping has activation character and may be written as follows:

$$\sigma = \sigma_2 \exp(-w_2/kT) \tag{3.15}$$

where pre-exponential factor $\sigma_2 \leq \sigma_1$ and activation energy of hops w_2 equals approximately half of localized states zone width in the middle of the gap ($w_2 = (E_1 - E_2)/2$).

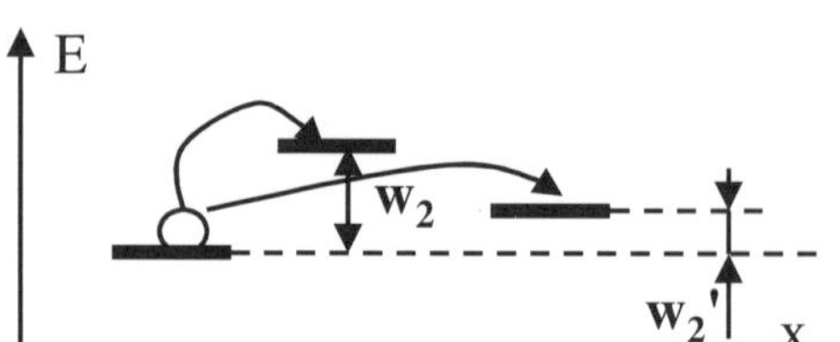

Figure 3.17. Schematic illustration of nearest — neighbor and variable — range hopping.

At low temperature the probability of nearest — neighbor hopping decreases. In this situation hopping between two sites whose transfer integral is smaller, but activation energy (w_2') is smaller as well becomes probable (Fig. 3.17). The conductivity temperature dependence for this case, given by Mott, is follows:

$$\sigma = \sigma_3 \exp[-(T_0/T)^{1/4}] \tag{3.16}$$

where σ_3 and T_0 are constants.

Total conductivity of material is determined as integral extended along all electron states. In the case of Boltzmann distribution of electrons above the Fermi level it is given by

$$\sigma = e \int N(E) f(E) \mu(E) dE \tag{3.17}$$

where $f(E)$ is Boltzmann function and e is electron charge. Energy dependencies of equation terms (3.17) are shown in Fig. 3.18. Boltzmann function is given for different temperatures $T_1 > T_2 > T_3$. At the temperature T_1 charge carriers are excited to delocalized states ($E > E_C$). Density of states $N(E)$ and carrier mobility $\mu(E)$ are large enough in this case and conductivity $\sigma(E)_{T1}$ is high as well. At lower temperature T_2 there are no carriers in delocalized states but there are carriers in the conduction band tail ($E_C > E > E_A$). However density of states and carrier mobility are significantly less in this case and engendered by these carriers conductivity ($\sigma(E)_{T2}$) is lower too. At lowermost temperature T_3 conductivity $\sigma(E)_{T3}$ is carried out by carriers at the states near the Fermi level.

The temperature dependence of non-crystalline semiconductor conductivity in a wide temperature range is shown in Fig. 3.19. There are four regions on this curve:

(1) at high temperature the conductivity in extended states is observed (Eqns. 3.7 and 3.8);
(2) next region is the hopping conduction in the band tails (Eqns. 3.13 and 3.14);
(3) the third region corresponds to nearest-neighbor hopping (Eqn. 3.15);
(4) and the last one is variable-range hopping (Eqn. 3.16).

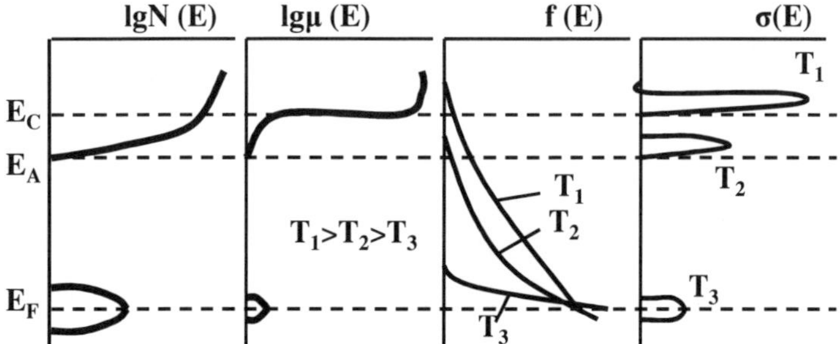

Figure 3.18. Energy dependencies of functions from Eqn. (3.16).

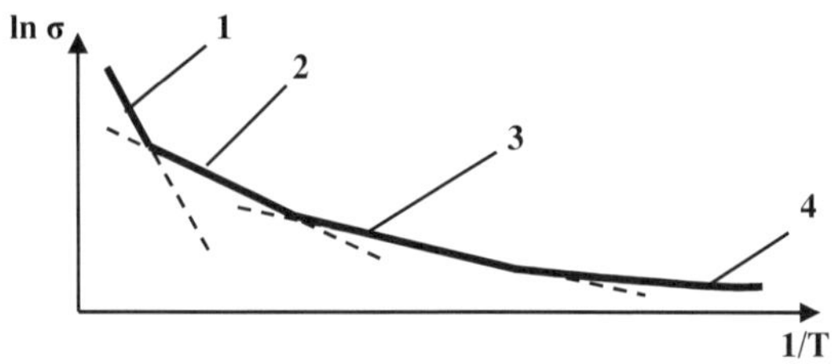

Figure 3.19. Schematic illustration of electrical conductivity temperature dependence of non-crystalline semiconductors (1 — conductivity in extended states; 2 — hopping conduction in the band tails; 3 — nearest-neighbor hopping and 4 — variable-range hopping).

The types of electron transport described above determine features of conductivity frequency dependence of non-crystalline semiconductors. In the case of carrier transport in extended states (energy more than E_C and less than E_V) frequency dependence of conductivity is described as follows:

$$\sigma(\omega) = \sigma(0)/(1 - \omega^2\tau^2) \tag{3.18}$$

where $\sigma(0)$ is direct current conductivity, ω is angular frequency and τ is relaxation time. Relaxation time in covalent bonded materials equals to 10^{-15}second approximately because of electronic nature of polarization. It means that there is no conductivity dependence on frequency in radiofrequency band in this case.

Logically, hopping conduction in the band tails (energy between E_C and E_A and between E_V and E_B) and in localized states near the Fermi level (energy between E_1 and E_2) has to depend on frequency In the case of direct current carrier has to pass through the sample from one electrode to the other. Direct current resistance is determined by the biggest barrier value between neighbor's localized states. Increase of frequency leads to a decrease in the number of localized states and barriers which carrier passes during half-cycle. This is way frequency dependence of hopping conductivity is given by

$$\sigma(\omega) = C\omega^{-S} \tag{3.19}$$

where C is constant, $S = 0.7$–1.0.

Frequency dependencies of conductivity in the case of different types of electron transport are shown in Fig. 3.20.

In practice, the type of electron transport providing the largest value of conductivity at a given temperature will dominate.

3.2.2 Thermoelectric Power and Hall Effect

Thermoelectric power (S) is determined as a ratio of voltage value (ΔU) between two points of semiconductor at different temperatures and the temperature difference (ΔT) of these points. In the case of n-type crystalline semiconductors thermoelectric power is given by

$$S = \Delta U/\Delta T = (k/e)[(E_C - E_F)/(kT) + A] \tag{3.20}$$

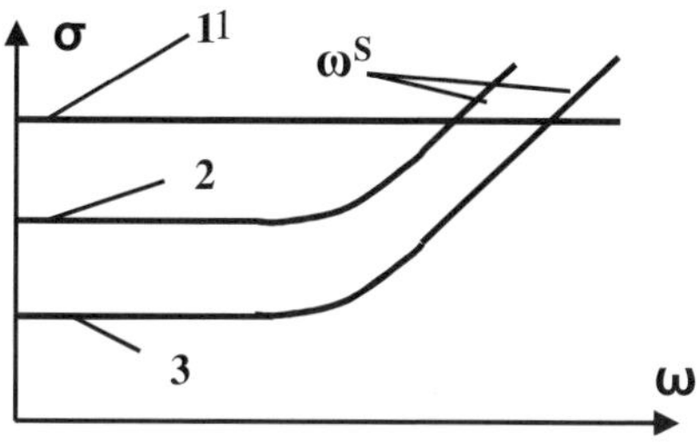

Figure 3.20. Schematic illustration of conductivity frequency dependences in the case of carrier transport in extended states (1), in band tails (2) and near the Fermi level (3).

where k is Boltzmann constant, A is coefficient depended on carrier scattering mechanism ($A = 2 - 4$ for crystalline semiconductors) and e is electron charge. Thermoelectric power of n-type semiconductors is negative because electron charge is negative. In the case of p-type crystal it is necessary to change $(E_C - E_F)$ to $(E_F - E_V)$ and to change the negative electron charge to positive hole charge in Eqn. (3.20). As a result thermoelectric power of p-type semiconductors is positive. Therefore the thermoelectric power polarity sign gives information about the conduction type of a semiconductor.

In the case of carrier transport in extended states of non-crystalline semiconductors the value of coefficient A in Eqn. (3.20) changes because carrier scattering increases as a result of long range order absence. As a rule $A = 1$ in this case and thermoelectric power for semiconductors of n and p-type are given by

$$S = (k/e)[(E_C - E_F)/(kT) + 1] \tag{3.21}$$

$$S = -(k/e)[(E_F - E_V)/(kT) + 1] \tag{3.22}$$

Transition to carrier transport in band tails leads to further lowering of coefficient A value ($A < 1$) owing to a decrease in the density of states. On the other hand $(E_C - E_F)/(kT) = 1$ and $(E_F - E_V)/(kT) = 1$ because the Fermi level lies near the middle of the gap. Therefore it is possible to ignore coefficient A. Besides, energies E_C and E_V in Eqns (3.21) and (3.22) have to change to E_A and E_B (Fig. 3.16). As a result we obtain for n and p-type conduction

$$S = (k/e)(E_A - E_F)/(kT) \tag{3.23}$$

$$S = -(k/e)(E_F - E_B)/(kT) \tag{3.24}$$

Equations (3.21–3.24) show that temperature dependency of thermoelectric power is determined by temperature dependency of carrier concentration both for carrier transport in extended states ($E_C - E_F$ and $E_F - E_V$) and for carrier transport in band tails ($E_A - E_F$ and $E_F - E_B$). In the case of electrical conductivity (Eqns. 3.7, 3.8, 3.13 and 3.14) this statement is correct only for carrier transport in extended states ($E_C - E_F$ and $E_F - E_V$). If carrier transport is in band tails, conductivity activation energy equals the sum of carrier concentration and carrier mobility activation energies ($E_F - E_V + w_1$ and $E_A - E_F + w_1$). Therefore it is possible to get information about the carrier transport mechanism from the comparison of conductivity and thermoelectric power temperature dependencies. If dependencies $\ln\sigma(1/T)$ and $S(1/T)$ have the same slope absolute values it means that carrier transport in extended states has taken place. In the case of different absolute values of slope carrier transport is in band tails and slope difference gives hop activation energy w_1.

In the case of hopping conductivity in localized states at the Fermi level thermoelectric power may be calculated by expression used for metals (because electrons move inside energy interval kT near the Fermi level in both cases):

$$S = \frac{\pi^2}{3} \cdot \frac{k^2 T}{e} \frac{d(ln\sigma)}{dE}\bigg|_{E=E_F} \tag{3.25}$$

Hall effect consists of electron trajectory deflection in transverse magnetic field due to Lorentz force and appearance of potential differences in the sample:

$$U = R_H(BI/d) \tag{3.26}$$

where B is magnetic induction, I is electric current, d is sample thickness and R_H is Hall coefficient

$$R_H = r/en \tag{3.27}$$

r is factor depending on scattering mechanism ($r \approx 1$), e is electron charge and n is carrier concentration. Hall effect measurement gives extensive information about electrical transport in crystalline semiconductors: type and concentration of charge carriers (Hall coefficient sign is negative for n-type and positive for p-type crystals).

However Hall effect measurement results show a number of anomalies caused by random structural network. Hall coefficient sigh is negative in chalcogenide glassy semiconductors in spite of these materials having p-type of conductivity. Moreover double reversal of R_H is observed in hydrogenated amorphous silicon. Hall coefficient is positive in phosphorus doped (n-type) a–Si:H and negative in boron doped (p-type) material. A number of models have been proposed to explain these anomalies (see for example Okamoto, Hattori and Hamakawa, 1993). But we just point out here that the measurement of the Hall effect is not a good tool for determination of carriers' charge sign in non-crystalline semiconductors.

3.2.3 Time of Flight Method

An effective method of determination of localized states density distribution in band tails of non-crystalline semiconductors is the drift mobility measurement by time of flight experiment. Diagram of the experiment is shown in Fig. 3.21. Voltage pulse is applied to the sample with two blocking electrodes (one of which is semi-transparent electrode).

As a result electric field is induced in the sample. The sample is exposed to a short pulse of strong absorbed light through semi-transparent electrode. Electrons and holes are created in the sample near semi-transparent electrode due to light

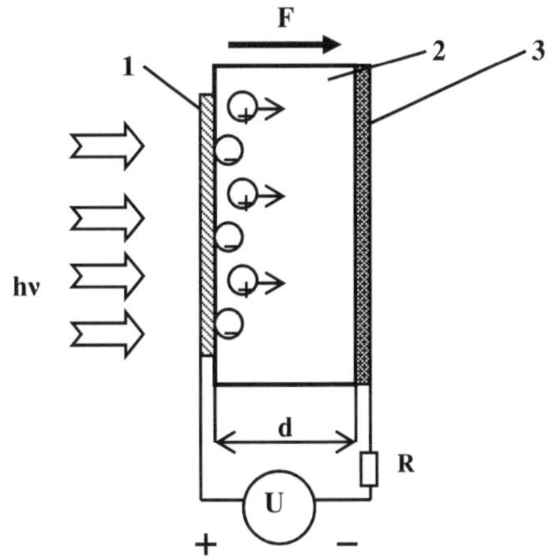

Figure 3.21. Diagram of time of flight experiment (1 is semi-transparent electrode, 2 is semiconductor film, 3 is electrode, d is film thickness and F is electric field intensity).

absorption. Positive or negative carrier packet (depending on voltage polarity) drifts across the sample under electric field action. Electric current is flowing in external circuit during carrier packet drift. Drift mobility is expressed by

$$\mu_D = d/(F \cdot t_D) \tag{3.28}$$

where d is sample thickness, F is electric field intensity and t_D is time of carrier flight. In ideal case of Gaussian conduction carrier packet reaches opposite electrode in time t_D and electric current in external circuit drops to zero sharply.

In the case of non-crystalline semiconductors localized states in band tail act as carrier traps which capture carriers. In turn some carriers excite back from traps to band. Probability of the last process depends on localized state energy and measurement temperature. The processes of carrier capture and release leads to spread of carrier packet. Time dependence of carrier concentration is expressed by

$$n_D(t) = n_D(0) \cdot \exp(-t/\tau_D) \tag{3.29}$$

where $n_D(0)$ is carrier concentration at $t = 0$ and τ_D is average carrier lifetime. Time dependence of drift current in this case is shown in Fig. 3.22(a). One may see that it is impossible to determine time of flight from this curve. However it has been shown by Pfister and Scher (1978) that in the case of log-log plot this dependence has two linear parts (Fig. 3.22(b)). Intersection of these parts gives time of flight value t_D. As a result carrier drift mobility may be determined from expression (3.28). On the other hand carrier drift mobility may be expressed by

$$\mu_D = \mu_\sigma (N_C/N_t) \exp[-(E_C - E_t)/kT] \tag{3.30}$$

where μ_σ is Hall mobility, N_C and N_t are density of states at E_C andE_t levels. Hence it is possible to determine energy position of localized states $(E_C - E_t)$ from the slope of ln $\mu_D(1/T)$ dependence. Point of intersection of the curve and

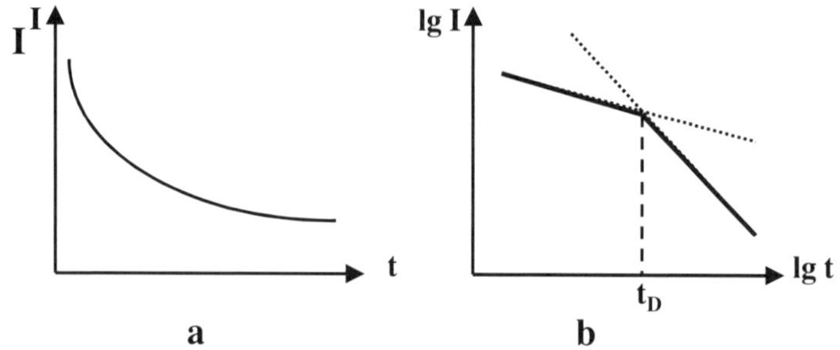

Figure 3.22. Schematic illustration of drift current dependence on time in the case of linear (a) and log-log (b) plots.

axis of ordinates gives the concentration of traps at the given energy. Therefore application of time of flight method in a wide temperature range gives information about localized state density distribution in band tail.

3.3 OPTICAL PROPERTIES OF DISORDERED SEMICONDUCTORS

In the vast spectrum of electromagnetic waves (from angstroms to kilometers) light or visible part of electromagnetic radiation takes up only a small part from 0.40 to 0.75 microns. It corresponds to frequencies $(7.5 \div 4.0) \bullet 10^{14}$ hertz and photon energies from 3.1 to 1.7 eV. Interaction of light flux (I_0) with homogeneous solid includes reflection (I_0R where R is reflection factor) and absorption (A). As a result light flux intensity may be written as follows:

$$I = I_0(1 - R)\exp[-\alpha(\bar{h}\omega)x] \tag{3.31}$$

where $\alpha(\bar{h}\omega)$ is absorption coefficient, $\bar{h}\omega$ is photon energy, $\bar{h}$ is Planck's constant and ω is frequency. Thus light flux is expressed by

$$I_0 = I_0R + I_0A + I_0T \tag{3.32}$$

where T is a part of flux transited through a sample, or

$$R + A + T = 1 \tag{3.33}$$

Interaction of light flux with solid may be described through refractive index as well. Complex refractive index is given by

$$n(\hbar\omega) = n(\hbar\omega) + ik(\hbar\omega) \tag{3.34}$$

where $n(\hbar\omega) = c/[v(\hbar\omega)]$ is real part of refractive index, $c = 3 \cdot 10^5\text{km/s}$ is light velocity in vacuum, v is light velocity in solid and $k(\hbar\omega)$ is imaginary part of refractive index. It correlates with absorption coefficient as follows:

$$\alpha(\hbar\omega) = 4\pi k(h\omega)/\lambda \tag{3.35}$$

where λ is wave length.

The main type of optical absorption in semiconductor materials is fundamental absorption. It leads to electron excitation from valence band into conduction band. The probability of photon absorption is determined by photon energy (it has to be no less than optical gap value $\hbar\omega \geq E_{opt}$) and by density of states distribution in valence band ($N(E)_V \sim E^p$) and in conduction band ($N(E)_C \sim E^q$). In the case of parabolic density of states $p = q = 1$ and dependence of absorption coefficient on photon energy is given by

$$\alpha n\hbar\omega = P(\hbar\omega - E_{opt})^m \tag{3.36}$$

where n is refractive index and P is constant. Index of a power $m = 1/2$ in the case of direct transitions and $m = 2$ in the case of indirect transitions (transitions that are accompanied by phonon emission). Indirect transitions predominate at the optical absorption edge when photon energy is close to gap value. Thus absorption coefficient of crystalline semiconductors depends on photon energy as square-law characteristic in this case. $(\alpha n\hbar\omega)^{1/2} - (\hbar\omega)$ plot yields a straight line extrapolation which, at its intersection with abscissa axis, gives the value of optical gap because $(\hbar\omega) = E_{opt}$ at $(\alpha n\hbar\omega)^{1/2} = 0$.

In contrast to crystals dependence of disordered semiconductor absorption coefficient on photon energy has four segments (Fig. 3.23). Portion A is the region of fundamental absorption. Portion B is the optical absorption edge where α depends on photon energy as square-low characteristic. These regions are similar to crystals. However in contrast to sharp subsequent decrease of absorption coefficient in crystals (dotted line in Fig. 3.23) square-low characteristic substitutes for exponential one in disordered semiconductors

$$\alpha = P\exp[-(\gamma/T^*)(E_{opt} - \hbar\omega)] \tag{3.37}$$

where P and γ are constants and parameter T^* is defined as follows:

$$T^* = T \text{ if } \quad T > T_{cr}$$
$$T^* = T_{cr} \text{ if } \quad T \leq T_{cr}$$

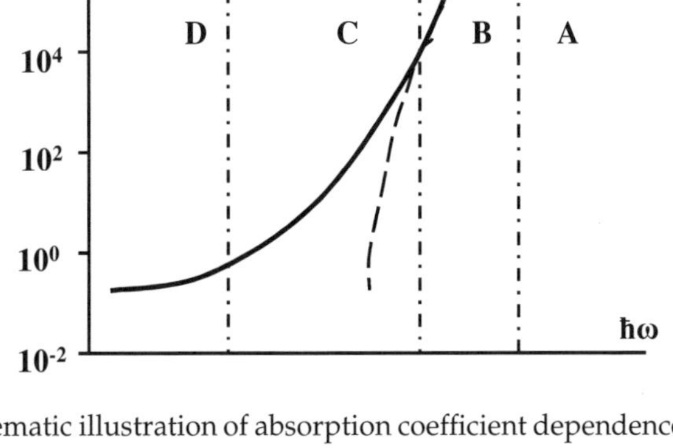

Figure 3.23. Schematic illustration of absorption coefficient dependence on photon energy.

where T_{cr} is some characteristic temperature of material. As a result exponential tail of absorption edge arises (portion C) which is called Urbach tail.

Exponential tail of absorption edge is generally observed in non-crystalline semiconductors and in some crystals as well (for example in crystals of trigonal selenium, cadmium sulphide and some others). The origin of Urbach tail does not have a single explanation. In the case of non-crystalline semiconductors it is tempting to explain Urbach tail by electron transitions between extended states of one band and localized states of another band tail due to exponential dependence of localized state density in the tail. But it was pointed out by Mott and Davis (1979) that this simple explanation is hardly probable. Optical absorption spectra of eight non-crystalline semiconductors at room temperature are shown in Fig. 3.24. As one can see from the figure curve slopes are approximately the same in the case of essentially different materials. It is difficult to assume that band tails are identical to such a degree in these materials. So there has to be a more general reason for exponential tail origin. As a rule Urbach tail is explained by the presence of exciton absorption or by the influence of internal electric field on absorption.

Portion D in the dependence of absorption coefficient on photon energy (Fig. 3.23) is caused by optical transitions between localized states in the middle of gap and extended states in conduction or valence band.

As mentioned above the value of optical gap in the case of crystalline semiconductors may be estimated by extrapolation of $(\alpha n\hbar\omega)^{1/2} - (\hbar\omega)$ plot to abscissa axis. It is possible to use the same procedure in the case of non-crystalline semiconductors. But absorption edge length (portion B) is rather small in this case due to the presence of Urbach tail. That is why extrapolation may produce a considerable error. If exponential tail is observed in optical absorption spectrum the value of optical gap may be found by so called Urbach focus point position.

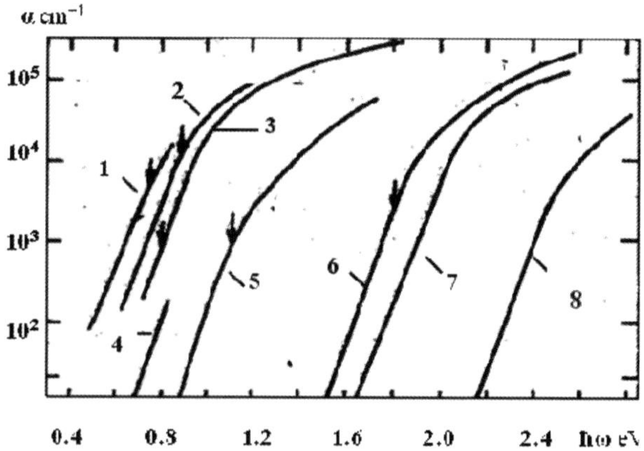

Figure 3.24. Optical absorption exponential tail of non-crystalline semiconductors at room temperature: 1-GeTe, 2-Te, 3-As_2Te_3, $4 - CdGeAs$, $5 - Ge_{16}As_{35}Te_{28}S_{21}$, $6 - As_2Se_3$, $7 - Se$, $8 - As_2S_3$. The arrows indicate doubled activation energies of electrical conductivity (Mott and Davis, 1979).

Schematic illustration of absorption coefficient spectral dependence in Urbach tail region at different temperatures is shown in Fig. 3.25. Taking into account the linear temperature dependence of gap value, Eqn. (3.37) at $T > T_{cr}$ may be given as follows

$$\alpha = P\exp[-\{\gamma(E_{opt}(0) - \hbar\omega - \beta T)/T\}] \tag{3.38}$$

where $E_{opt}(0)$ is optical gap value at 0 K and β is temperature coefficient of gap value. In the case of $\hbar\omega = E_{opt}(0)$ absorption coefficient ($\alpha = P\exp(\gamma\beta)$) does not depend on temperature (focus point A in Fig. 3.25) and position of this point gives the value of optical gap. This method is applicable to crystals and amorphous silicon. But in the case of chalcogenide glassy semiconductors characteristic temperature T_{cr} as a rule is more than glass transition temperature T_g. As a result Eqn. (3.37) transforms to

$$\alpha = P\exp[-\delta(E_{opt} - \hbar\omega)] \tag{3.39}$$

where $\delta = \gamma/T_{cr}$ is constant and absorption coefficient spectrum in Urbach tail region does not depend on temperature for solid materials. This does not allow to find Urbach focus point and value of optical gap.

There is another method to estimate optical gap value in non-crystalline semiconductors. It is based on determination of optical absorption coefficient value for photon energy equaling value of mobility gap. The latter is determined by electrical measurements. It is known that the Fermi level lies nearly in the middle of mobility gap in non-crystalline semiconductors. Consequently activation energy of conductivity in extended states is equal to approximately half of mobility gap value ($E_C - E_F = E_\sigma$). It means that optical gap value has to correspond to photon energy equal to double conductivity activation energy ($\hbar\omega = 2E_\sigma$). Optical absorption coefficient at room temperature is equal to approximately 10^3 cm^{-1} at

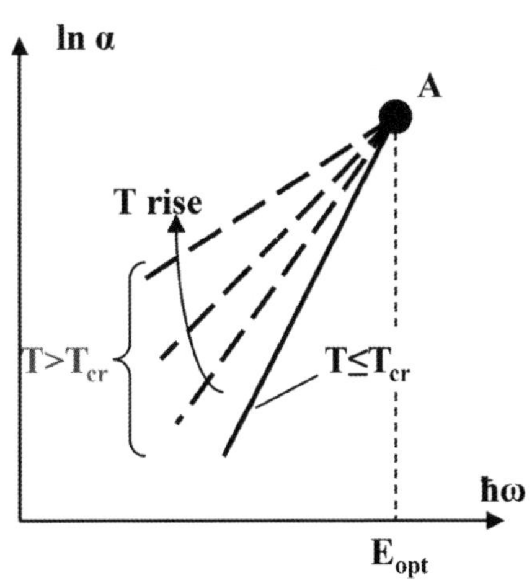

Figure 3.25. Schematic illustration of absorption coefficient spectral dependence in Urbach tail region at different temperatures (A is Urbach focus point).

this photon energy for many non-crystalline semiconductors (Fig. 3.24). So value of optical gap is determined as photon energy corresponding to optical absorption coefficient $\alpha = 10^3$ cm^{-1}. One can see that this method is rather imprecise as well.

3.4 PHOTOELECTRICAL PROPERTIES OF DISORDERED SEMICONDUCTORS

Photoelectrical properties of semiconductors are based on electric conductivity changes under the direct influence of light. If photon energy corresponds to regions A or B of absorption coefficient spectral dependence (Fig. 3.23) absorption of photons leads to electron — hole pairs generation. As a result charge carrier concentration and electric conductivity of the semiconductor increases. Total electric conductivity of semiconductor under illumination is given as follows:

$$\sigma = \sigma_d + \sigma_{ph} = e\mu_n(n + \Delta n) + e\mu_p(p + \Delta p) \tag{3.40}$$

where $\sigma_d = e\mu_n n + e\mu_p p$ is dark conductivity, $\sigma_{ph} = e\mu_n \Delta n + e\mu_p \Delta p$ is photoconductivity, Δn and Δp are excess charge carrier concentrations.

Time dependence of electron concentration is determined by continuity equation

$$\frac{\partial n}{\partial t} = -div\left(\frac{j_n}{e}\right) + G_n - \frac{\Delta n}{\tau_n} \tag{3.41}$$

where G_n is carrier generation rate and j_n is electron current density. The right part of Eqn. (3.41) shows electron concentration changes owing to electric current (the first term), charge carrier generation (the second term) and charge carrier recombination (the third term). Where electric current equals zero ($j_n = 0$) Eqn. (3.41) transforms to

$$\frac{\partial n}{\partial t} = G_n - \frac{\Delta n}{\tau_n} \tag{3.42}$$

And in the case of constant conditions ($\frac{\partial n}{\partial t} = 0$) Eqn. (3.42) transforms to:

$$G_n = \Delta n / \tau_n \tag{3.43}$$

Excess charge carrier concentration may be written as follows:

$$\Delta n = G_n \tau_n \tag{3.44}$$

Photocarrier generation rate is expressed by

$$G_n = I\alpha\eta \tag{3.45}$$

where $I = I_0(1 - R)$ is light flux inside the sample, I_0 is incident light flux, α is absorption coefficient and η is quantum efficiency. Quantum efficiency is determined as a ratio between the number of generated photocarriers and the number of absorbed photons and has value less one usually. Taking into account Eqn. (3.45) excess charge carrier concentration is given by

$$\Delta n = I\alpha\eta\tau \tag{3.46}$$

and photocurrent density is expressed as follows:

$$j_{ph} = e\Delta n\mu E = eI\alpha\eta\tau\mu E \tag{3.47}$$

where μ is carrier mobility and E is electric field intensity. On the other hand photocurrent density may be expressed from Ohm's law by

$$j_{ph} = \sigma_{ph}E \tag{3.48}$$

Photoconductivity may be determined from Eqns. (3.47) and (3.48) as follows:

$$\sigma_{ph} = eI\alpha\eta\tau\mu \tag{3.49}$$

As one can see from Eqn. (3.49) photoconductivity depends on four factors: light flux intensity ($\sigma_{ph} = f(I)$), light flux spectral characteristic ($\sigma_{ph} = f(\alpha), \alpha = f(\hbar\omega)$), temperature ($\sigma_{ph} = f(\alpha\mu\tau), \alpha, \mu, \tau = f(T)$) and electric field intensity ($\sigma_{ph} = f(\eta), \eta = f(E)$). Let us examine these dependencies.

Dependence of photoconductivity on light flux intensity. It has been shown from experiment that in the case of relatively small light flux intensity photoconductivity varies directly with it in accordance with Eqn. (3.49). But at high flux intensity photoconductivity rise becomes slower and the dependence is given by

$$\sigma_{ph} = AI^n \tag{3.50}$$

where A is constant and $n = 0.5 - 1.0$. Photocurrent dependence on light flux intensity is shown in Fig. 3.26. The reason for deviation from linear dependence is in the increase of excess charge carrier concentration. At high flux intensity it becomes considerably higher in comparison with equilibrium carrier concentration ($\Delta n >> n$). It leads to a rise of photocarrier recombination rate and to a decrease in their lifetime. As a result the lifetime in Eqn. (3.49) becomes a function of flux intensity too ($\tau = f(I)$).

Dependence of photoconductivity on photon energy and spectral dependence of absorption coefficient are shown in Fig. 3.27. One can see from the figure that

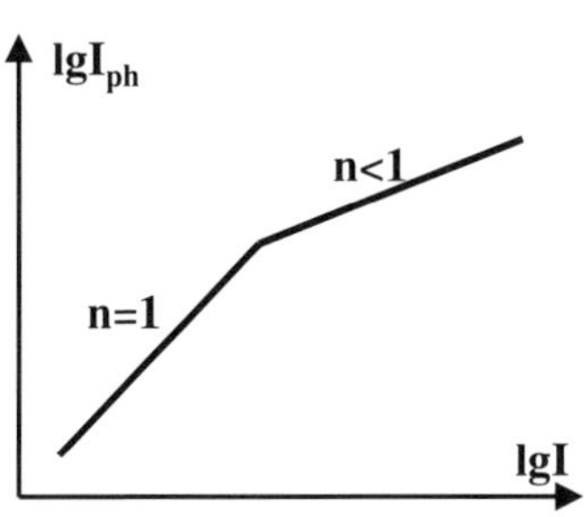

Figure 3.26. Schematic illustration of photocurrent dependence on light flux intensity.

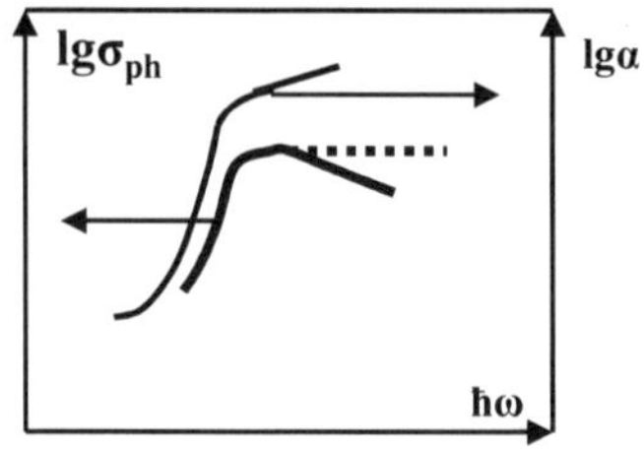

Figure 3.27. Schematic illustration of photoconductivity and absorption coefficient spectral dependences.

the dependence $lg\sigma_{ph}(\hbar\omega)$ has two considerably different parts. In the region of optical absorption edge photoconductivity repeats the spectral dependence of absorption coefficient in accordance with equation (3. 49). However all photons are absorbed in the sample at a certain value of absorption coefficient and further rise of photon energy does not lead to photoconductivity augmentation in spite of absorption coefficient increase. It is possible to assume that photoconductivity has to be constant in this region of photon energy (dotted line in the Fig. 3.27). But experiment indicates reduction of photoconductivity.

The photoconductivity decrease is connected with surface effects. Increasing of absorption coefficient value results in photon absorption free path decrease so absorption occurs nearer to the surface which has higher concentration of defect states. It leads to a rise of photocarrier recombination rate. The effect is observed both in crystalline and in non-crystalline semiconductors and it is more apparent in crystals because of lesser concentration of defect states in the volume.

Temperature dependence of photoconductivity is shown in Fig. 3.28. One can see that the dependence has two different regions. At low temperatures when $\sigma_{ph} > \sigma_d$ photoconductivity increases with rising temperature (region I). However at high temperatures when dark conductivity exceeds photoconductivity ($\sigma_{ph} < \sigma_d$) the latter decreases with rising temperature (region II).

There are three temperature dependent terms in equation (3.49): optical absorption coefficient (α), carrier lifetime (τ) and carrier mobility (μ). It is necessary to note that carrier drift mobility (μ_D, expression 3.28) has to be analyzed in

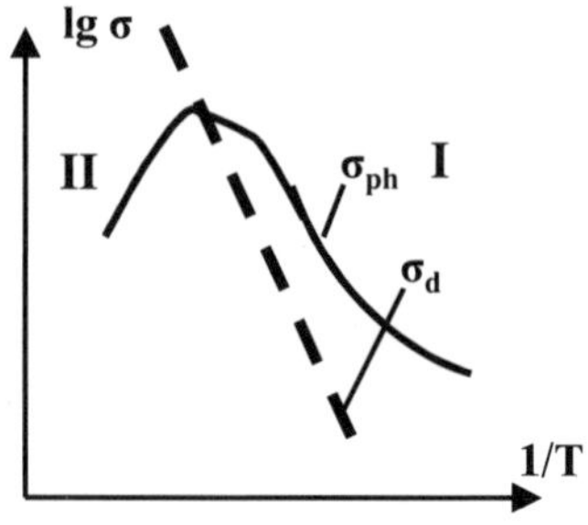

Figure 3.28. Schematic illustration of photoconductivity temperature dependence of non-crystalline semiconductors.

presence of a strong electric field. Among the characteristics mentioned above, carrier drift mobility has the strongest temperature dependence. It is determined by trapping processes of photocarriers on band tail localized states and it increases with rising temperature. In the case of $\sigma_{ph} > \sigma_d$ recombination of photoexcited electrons and holes takes place through deep recombination levels and photoconductivity temperature dependence is determined by temperature dependence of drift mobility. However when dark conductivity exceeds photoconductivity ($\sigma_{ph} < \sigma_d$) equilibrium carrier concentration exceeds concentration of photocarriers and recombination process between photoexited and thermoexited carriers takes place. Equilibrium carrier concentration increases with rising temperature. It leads to decrease in the photocarrier lifetime (τ) and to decrease in photoconductivity in accordance with Eqn. (3.49).

Electric field intensity dependence of photoconductivity is determined by quantum efficiency (η) dependence on electric field. Because quantum efficiency is a ratio between the number of generated photocarriers and the number of absorbed photons and η *is* usually < 1 the following question arises: why is the number of created electron — hole pairs contributing to photoconductivity is less than the number of absorbed photons? To answer this question let us examine the process of optical generation of the electron — hole pair.

Absorption of photon with energy $\bar{h}\omega \geq E_{opt}$ leads to electron transition from valence to conduction band and creation of electron — hole pair with energy difference $E \approx \bar{h}\omega$ (Fig. 3.29). There are two possible ways of subsequent development. Electron and hole may separate under the influence of electric field or concentration gradient (solid arrows in the figure) and become free charge carriers. Alternatively, charge carriers may lose part of their energy by phonon scattering, couple through Coulomb attractive interaction and recombine with each other (double arrow in the figure).

Because electric field is conducive to charge carrier separation, an increase in electric field intensity leads to quantum efficiency rise. The value of quantum efficiency becomes close to one in the case of electric field intensity of 10^5 V/cm (Fig. 3.30, solid line). This explains experimental electric field dependence of photoconductivity usually observed in non-crystalline semiconductors. However it has been shown by Juska, Arlauskas and Montrimas (1987) that quantum

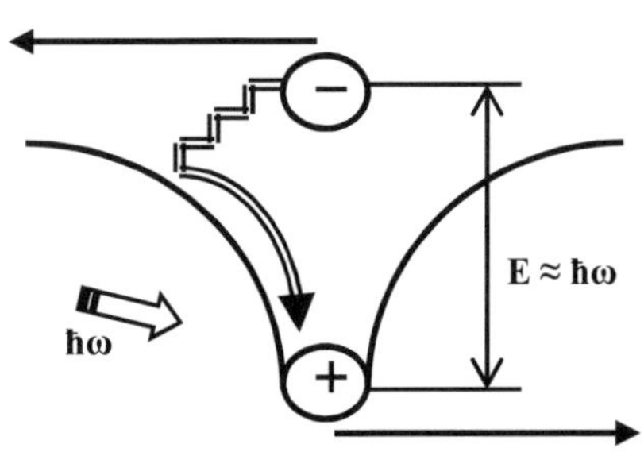

Figure 3.29. Illustration of optical generation and recombination processes (see text).

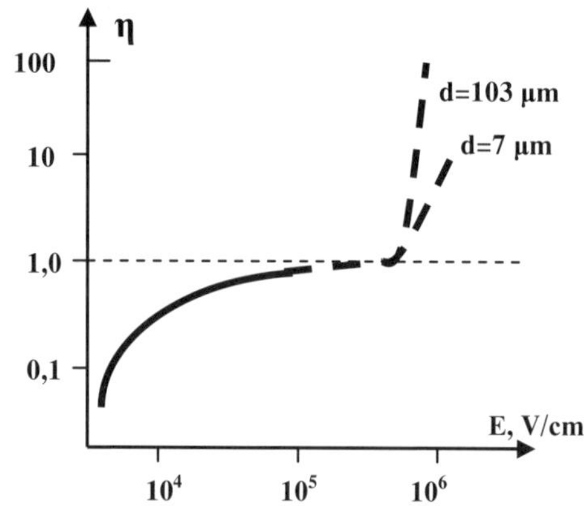

Figure 3.30. Field dependence of quantum efficiency in non-crystalline selenium (d is film thickness).

efficiency in non-crystalline selenium films exceeds one in the case of electric field intensity of more than 10^5 V/cm (Fig. 3.30, dashed lines). In other words the absorption of one photon leads to formation of several (or several dozens) electron — hole pairs. This phenomenon may be explained by avalanche carrier multiplication in electric field. In this case increase in film thickness at constant electric field intensity has to produce a rise in quantum efficiency because the greater number of carrier ionization acts. As one can see from the figure, the experiment confirms this assumption: quantum efficiency value increases sharply in the thick selenium films.

The existence of avalanche carrier multiplication in non-crystalline semiconductor is a rather unexpected result because the absence of long range order in atomic arrangement leads to small carrier free path. However it allowed to increase sensitivity of TV camera tubes (vidicons) by dozens times (see Chapter 6).

Chapter Four

Methods for Controlling Properties of Disordered Semiconductors

A distinctive feature of crystalline semiconductors is the strong dependence of electrical conductivity on the presence of impurities. Admixture of a thousandth of a fraction per cent impurity leads to the change of electrical conductivity by orders of magnitude and to the change of conductivity type. It is precisely this availability of a simple electrical property control method by doping that ensures the progress of modern solid state electronics and microelectronics.

Excess electron of a donor (phosphor atom in silicon for example) transfers to anti-bonding orbital and forms a donor center under conduction band bottom in the case of donor doping. It leads to a shift of the Fermi level towards conduction band and to an increase in free electron concentration. Similar situation occurs in the case of acceptor doping. However, doping of non-crystalline semiconductors does not lead to noticeable changes in their electrical properties. The insensitivity of non-crystalline semiconductors to impurities has been discovered by Kolomiets (1960) and has become a characteristic feature of these materials.

Let us consider the reasons for this phenomenon. In the case of chalcogenide glassy materials with flexible disordered structural network each foreign atom can satisfy its valency requirements and will not act as an electrically active center. Even if a number of foreign atoms remains electrically active, their impact is negligible. Such effect originates in relaxation of structural network that leads to a variation in the ratio between negatively (C_1^-) and positively (C_3^+) charged intrinsic defect states (VAP states, see Chapter 3). Concentrations of these states are same in the undoped material (Fig. 4.1(a)). A rise in concentration of electrically active impurity centers (donors for example) leads to electron redistribution between intrinsic and extrinsic defect states. Concentration of C_3^+ centers decreases and concentration of C_1^- centers increases (Fig. 4.1(b)). As a result the Fermi level position and electrical conductivity remain unchanged.

Aforementioned reasoning cannot be applied to amorphous silicon which has rigid four-fold coordinated structural network. A rigid structural network does

Disordered Semiconductors: Physics and Applications by Anatoly Popov

www.panstanford.com
978-981-4241-76-2

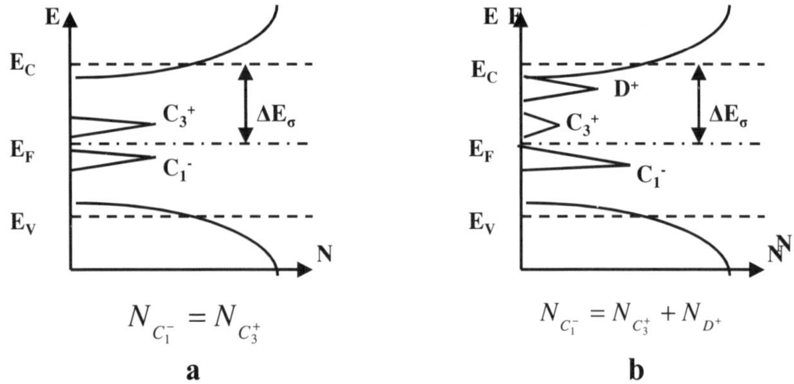

Figure 4.1. Density of charged states in the gap of undoped (a) and doped (b) chalcogenide glassy semiconductor (ΔE_σ is conductivity activation energy).

not allow easy satisfaction of the valency requirements for foreign atoms. On the other hand rigid structural network leads to high concentration of intrinsic defects. The concentration of foreign atoms proved to be less than intrinsic defect concentration (Fig. 3.15(a)) and impurities do not influence the Fermi level position and electrical conductivity value.

The insensitivity of non-crystalline semiconductors to impurities creates huge difficulties in device design. That is why some properties control methods for these materials have been proposed:

- doping of hydrogenated amorphous silicon,
- chemical modification of chalcogenide films,
- inversion of conductivity type in balk glassy chalcogenide,
- structural modification of non-crystalline semiconductors.

4.1 DOPING OF HYDROGENATED AMORPHOUS SILICON

The possibility of purposeful material property changes by means of doping is the feature that distinguishes hydrogenated amorphous silicon from other non-crystalline semiconductors. This possibility was demonstrated by Spear and Le Comber in 1975 and allowed a–Si:H wide practical application observed today.

As mentioned in Chapter 3, incorporation of hydrogen atoms in amorphous silicon sharply decreases concentration of intrinsic defects and density of localized states in the gap. As a result, donor (or acceptor) level density exceeds density of intrinsic defect levels (Fig. 3.15(a)) after doping. It leads to a shift in the Fermi level position and allows to change electrical conductivity value by up to 6–10 orders of magnitude and to change type of conductivity as well.

Because generally used method of amorphous hydrogenated silicon films preparation is silane (SiH_4) glow discharge decomposition (see Chapter 5), doping is carried out by addition of gases containing donor or acceptor atoms to silane.

Gas mixture of silane, hydrogen and phosphine (PH_3) is used in the case of phosphorus doping and gas mixture of silane, hydrogen and diborane (B_2H_6) is used in the case of boron doping. As a rule doping level is controlled by changing of ratio PH_3/SiH_4 or B_2H_6/SiH_4 in the range from 10^{-7} to 10^{-1}.

Dependences of a–Si:H electrical conductivity at room temperature and activation energy of conductivity on level of doping by phosphorus and boron are shown in Fig. 4.2 (Spear and Le Comber, 1976). Conductivity of undoped a–Si:H equals $10^{-8} - 10^{-9}$ $\text{Ohm}^{-1}\text{cm}^{-1}$. Increasing phosphorus concentration leads to smooth conductivity rise up to 10^{-2} $\text{Ohm}^{-1}\text{cm}^{-1}$. Activation energy of conductivity decreases from 0.6 to 0.2 eV simultaneously. These results are indicative of the Fermi level shift to conduction band. It is necessary to note that small phosphine additives to silane $PH_3/SiH_4 = 10^{-5} - 10^{-4}$) lead to a sharp conductivity increase in comparison with undoped material. It indicates relatively low localized state

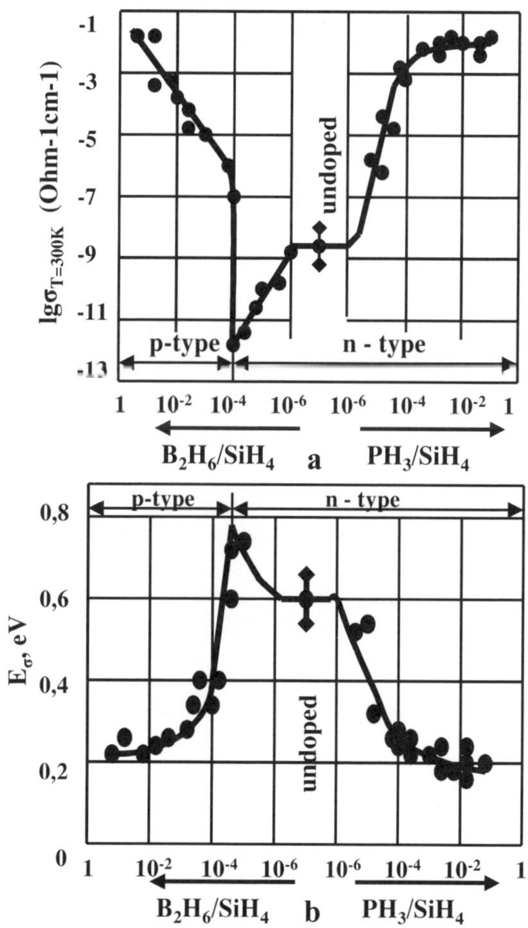

Figure 4.2. Dependences of electrical conductivity (a) and activation energy of conductivity (b) a-Si:H on doping by phosphorus and boron level (Spear and Le Comber, 1976).

density in the middle of gap. At higher levels of phosphine content the Fermi level draws nearer to conduction band. Movement of the Fermi level slows down because high localized state density in conduction band tail and conductivity saturation is observed at $\sigma_{T=300K} \approx 10^{-2}\text{Ohm}^{-1}\text{cm}^{-1}$and $E_\sigma \approx 0.2$ eV.

In the case of boron admixture the situation is different. Small diborane additives to silane lead to a decrease in conductivity of a–Si:H by four order of magnitude to the value 10^{-12} $\text{Ohm}^{-1}\text{cm}^{-1}$. Activation energy of conductivity increases to 0.8 eV. When gas ratio B_2H_6/SiH_4 exceeds the value of $10^{-5} - 10^{-4}$ conductivity begins to rise together with diborane concentration increase. The maximum conductivity value is approximately the same as in the case of phosphorus doping. Conductivity rise is accompanied by activation energy of conductivity decrease from 0.8 to 0.2 eV.

The dependencies described above may be explained under the assumption that undoped hydrogenated amorphous silicon is not an intrinsic semiconductor. It has weak n–type conductivity and the Fermi level lies a little bit above the middle of gap. In the case of phosphorus doping, the shift of the Fermi level up to conduction band leads to a decrease of activation energy of conductivity and to an increase of conductivity. In the case of boron doping, the Fermi level shifts down to valence band but it leads to an increase in activation energy of conductivity and to a decrease of conductivity because the n-type of conductivity still remains. At the gas ratio of approximately 10^{-4} the Fermi level reaches the middle of gap and conductivity activation energy gains the maximum value. The increase in diborane concentration after this point yields to p-type conductivity and to lower activation energy of conductivity.

Doping efficiency depends on film preparation conditions. However, comparison of phosphorus (or boron) atom concentration with the value of electrical conductivity of the film shows that only one third of admixture atoms affect the Fermi level position at best (Spear and Le Comber, 1984). There are two reasons for this fact. Firstly, only part of phosphorus (boron) atoms is four-fold coordinated and acts as donors (acceptors) in a disordered structural network. It has been shown (Stutzmann *et al.*, 1987) that concentration of four-fold coordinated phosphorus atoms is proportional to the square root of the total incorporated atom concentration. Secondly, doping mechanism in hydrogenated amorphous silicon differs from that in a single crystal. In the former case, excess electrons of donor atoms occupy empty localized states near the Fermi level. It leads to a shift of the Fermi level towards conduction band. Value of the shift (ΔE_F) depends on the ratio of donor concentration (N_D) to localized state density near the Fermi level ($g(E_F)$):

$$\Delta \mathrm{E_F} \sim \mathrm{N_D}/g(\mathrm{E_F}) \tag{4.1}$$

In the case of fixed N_D the value of the Fermi level shift is inversely proportional to $g(E_F)$. Because localized state density near the Fermi level in non-hydrogenated amorphous silicon reaches 10^{20} cm^{-3} it is clear why doping does not influence the Fermi level position and properties of this material.

Positions of phosphorus donor level and boron acceptor level in hydrogenated amorphous silicon are $E_C - 0.2$ eV and $E_V + 0.3$ eV respectively.

4.2 CHEMICAL MODIFICATION OF CHALCOGENIDE GLASSY SEMICONDUCTOR FILMS

As mentioned above, chalcogenide glassy semiconductors are not sensitive to impurities because a foreign atom can satisfy its valency requirements and it does not act as an electrically active center due to relaxation of structural network in a material lacking long-range order. Therefore it is necessary to prevent structural relaxation during material preparation in order that foreign atoms remain electrically active centers. Thus, from the point of view of physics, the idea of chalcogenide glassy films chemical modification consists in incorporation of impurity atoms into growing chalcogenide film under conditions where atomic structure relaxation is not probable. These conditions exist below glass transition temperature (Ovshinsky, 1977).

This idea was realized by simultaneous radio-frequency sputtering of chalcogenide glass and the desired amount of additive. Additive concentration in the film is varied by changing the number of small additive disks attached to the face of the sputtering target of the host material (Flasck *et al*, 1977). Conductivity temperature dependences of $Ge_{32}Te_{32}Se_{32}As_4$ thin films modified by nickel are shown in Fig. 4.3. Electrical characteristics of films with various nickel content are listed in Table 4.1.

As one can see from the figure and the table, conductivity activation energy of unmodified film is constant throughout the temperature range and equals

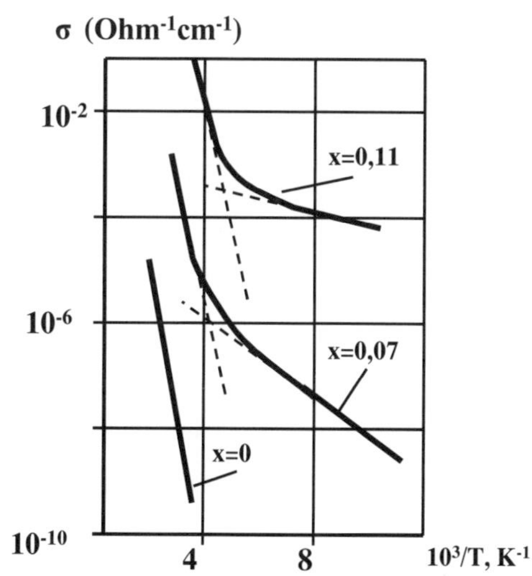

Figure 4.3. Conductivity dependences of$(Ge_{32}Te_{32}Se_{32}As_4)_{1-x}Ni_x$ films on temperature.

Table 4.1 Characteristics of chemically modified films $(Ge_{32}Te_{32}Se_{32}As_4)_{100-x}Ni_x$.

X, at %	$E_\sigma ev$	$\sigma_{293K} hom^{-1}cm^{-1}$	S, μV/K
0	0.74	$4 \cdot 10^{-10}$	+2000
7	0.33	$1.4 \cdot 10^{-5}$	+54
9.1	0.25	$6.5 \cdot 10^{-4}$	−82
11.4	0.18	$3.8 \cdot 10^{-2}$	−84

approximately half of chalcogenide glass gap value. Nickel modification leads to an increase in conductivity value by up to eight orders of magnitude. Beside that, the region of extrinsic conduction appears on the $\sigma(1/T)$ curves. Rise of nickel concentration results in a decrease in conductivity activation energy from 0.74 to 0.18 eV. At the same time, the value of optical gap changes only insignificantly. In addition this modification causes changes of thermoelectric power (Seebeck coefficient) from positive to negative value (table 4.1). It means that transition from p-type conductivity to n-type conductivity of material takes place. These results confirm that the reason for electrical characteristic changes is a shift of the Fermi level. Similar effects have been observed with different transition metal modifiers in a number of chalcogenide glasses (Flasck *et al*, 1977).

Thus chemical modification of chalcogenide glassy films leads to effects similar to doping of crystalline semiconductors and hydrogenated amorphous silicon.

4.3 CONDUCTIVITY TYPE INVERSION IN BULK GLASSY CHALCOGENIDE

Results of thermoelectric power measurements indicate p–type conductivity for all bulk chalcogenide glasses. Possibility of changing conductivity type in chalcogenide glassy films by chemical modification method (see Sec. 4.2) gives rise to attempts to prepare bulk chalcogenide glass with n–type conductivity. Usually it is done by adding metals to the melt before quenching. A lot of materials were tested. However in the case of small amount of added metal (within the limits of a few per cent) glasses retain p–type conductivity. Increase of metal content in the melt leads to crystallization of material during cooling. For example, concentration of less than one atomic per cent of nickel or chrome can be realized in glassy As_2Se_3 without crystallization of material. In the case of glassy selenium highest possible bismuth concentration is two atomic per cent. Substitution of selenium for system As–Se allows to increase bismuth concentration to up to 4 atomic per cent but the material retains p–type conductivity.

The largest amount of metal can exist in glasses of Ge–S and Ge–Se systems. Glass formation regions for Ge–Se–Bi system at various cooling rates are shown

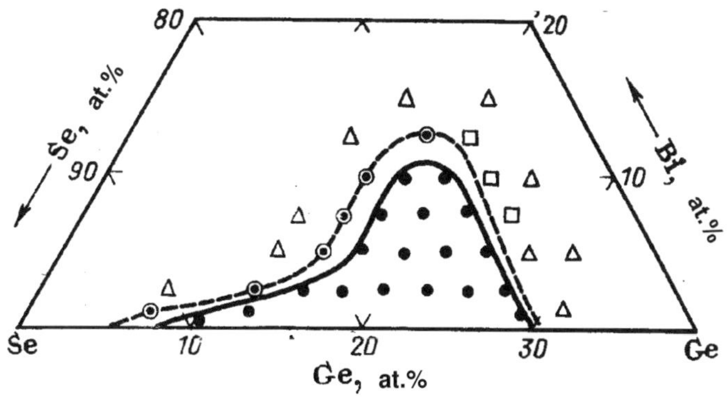

Figure 4.4. Glass formation regions for Ge–Se–Bi system (•–slow cooling, ⊡–cooling in ice water, □–glassy-crystalline phase, △–crystalline phase).

in Fig. 4.4 (Pazin and Borisova, 1969). It is possible to introduce up to 14 atomic per cent of bismuth to the glasses $GeSe_{3.5}$ and $GeS_{3.5}$ without crystallization of the material.

Electrical conductivity temperature dependencies of bulk glasses $(GeSe_{3.5})_{100-x}Bi_x$ are shown in Fig. 4.5 (Aivasov *et al.*, 1995). Increase of bismuth concentration leads to a rise of absolute conductivity values by up to six orders of magnitude and to a decrease in conductivity activation energy (the slope of curves $\lg\sigma(1/T)$ decreases). Beside that, Seebeck coefficient (thermoelectric power) changes from positive to negative sign at bismuth content of about 8 atomic per cent. This indicates change of conductivity from p–type to n–type. As a rule these facts are interpreted as the Fermi level shift towards the conduction band.

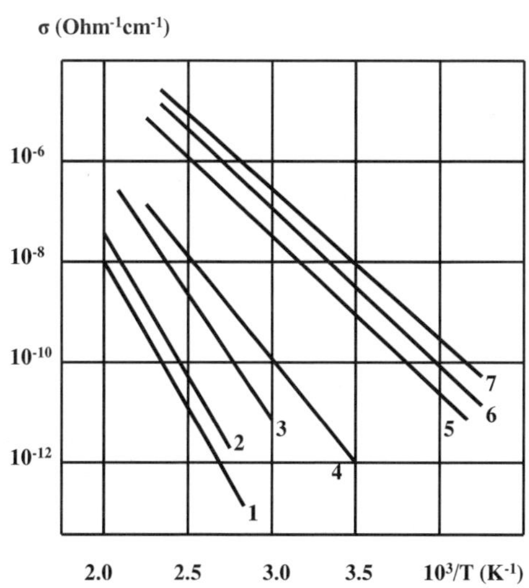

Figure 4.5. Conductivity temperature dependence of $(GeSe_{3.5})_{1-x}Bi_x$ glasses for: $x = 0(1), x = 0.04(2), x = 0.06(3), x = 0.08(4), x = 0.10(5), x = 0.12(6)$ and $x = 0.14(7)$.

However, in contrast to chemical modification of glassy chalcogenide films (see Sec. 4.2) there is no region of extrinsic conduction on the conductivity temperature dependence curves in this case. Dependences $\lg \sigma\ (1/T)$ are straight lines with constant slope throughout the temperature range. It contradicts the assumption about the Fermi level shift. Increase in the absolute value of conductivity and decrease in conductivity activation energy may arise from both the Fermi level shift and from the change of forbidden gap value. Measurements of optical gap value in the $(GeSe_{3.5})_{1-x}Bi_x$ glasses show that optical gap value decreases from 1.95 to 1.20 eV when bismuth concentration rises from 0 to 14 atomic per cent. Moreover, concentration dependences of conductivity activation energy and half of optical gap have analogous characteristics and similar absolute values (Fig. 4.6, Aivasov *et al.*, 1995). It indicates that the observed changes of electrical properties are due to value of forbidden gap lowering without the Fermi level shift.

However it is difficult to explain changes of Seebeck coefficient sign and conduction type in samples with bismuth content eight and more atomic per cent based on this assumption. Is the change of conduction type possible when the Fermi level position does not change?

It is known that hole conduction is equal to electron conduction in the case of intrinsic semiconductor. Because conductivity $\sigma = en\mu_n + ep\mu_p$ the condition of intrinsic conduction is equality $n\mu_n = p\mu_p$. Hole mobility in chalcogenide glassy semiconductors exceeds electron mobility. It means that $p\mu_p > n\mu_n$ and thermoelectric power has positive sign even if the Fermi level is situated at the middle of gap. Bismuth creates deep traps for holes in chalcogenide glasses but has weak influence on electron mobility. As a result hole mobility becomes less than electron mobility at certain bismuth concentration and sign of thermoelectric power changes in spite of constant position of the Fermi level.

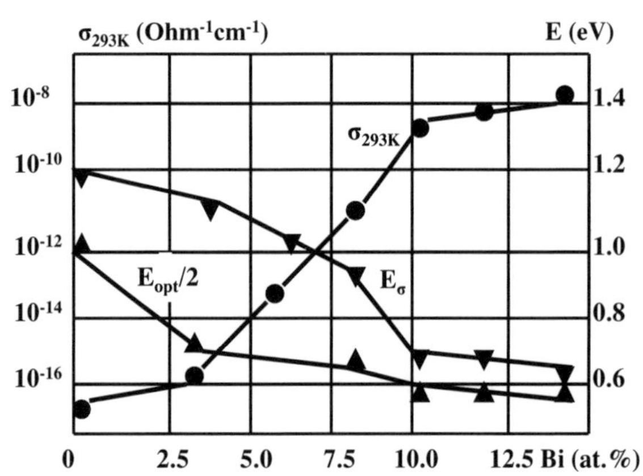

Figure 4.6. Dependencies of room temperature conductivity (σ_{293K}), activation energy of conductivity (E_σ) and half of optical gap value ($E_{opt}/2$) on bismuth concentration in the glasses $(GeSe_{3.5})_{1-x}Bi_x$.

4.4 STRUCTURAL MODIFICATION OF DISORDERED SEMICONDUCTORS PROPERTIES

As mentioned above the most essential feature of disordered semiconductors is their weak sensitivity to impurities. The problems of control over the properties and reproducible synthesis of these materials with prescribed properties still remain pressing in spite of some solutions being found (see Sec. 4.1–4.3).

As an alternative to semiconductor properties control by doping, the method of structural modification of disordered semiconductors properties has been proposed. The method allows to control properties by changing the material's structure without changing its chemical composition (Popov, Michalev and Shemetova, 1983). The physical basis for the method is the fact that free energy of disordered materials has several minima corresponding to various metastable states of the system (see Chapter 1).

It has been shown in Chapter 1 that there are four levels of structural characteristics necessary for common description of structure of disordered solids, namely: short range order in atomic disposition, medium range order in atomic disposition, morphology and subsystem of defects. Structural changes are possible at all of the levels of structural characteristics mentioned above. So one can conclude that, on the whole, there are four levels of structural modification that differ by type of change of material structure namely: level of short range order (level 1), level of medium range order (level 2), level of morphology (level 3) and level of defect subsystem (level 4) (Table 4.2, Popov, Vorontsov and Popov, 2001).

4.4.1 *Structural Modification at the Level of Short Range Order*

Structural changes at the level of short-range order lead to variations of all basic properties of material. For example, polymorphic crystalline modifications of carbon (diamond, graphite and carbine) possess fundamentally different physicochemical properties because of the different hybridizations of electron orbitals and different atomic structures at the short-range order level. Amorphous carbon films incorporate structural units of different allotropic modifications, with the relative content of these units determined by film growth modes and varying widely for the same preparation method. Correspondingly, the coordination of atoms varies (between 2 and 4) together with other parameters of the first coordination sphere (Vasil'eva and Popov, 1995). When films of amorphous hydrogenated carbon (a–C:H) are obtained by rf-ion-plasma sputtering in an argon-hydrogen atmosphere, merely changing the substrate temperature and discharge power may give films (Popov*et al.*, 1995), in which the optical gap varies by two orders of magnitude (between 0.02 eV for graphite-like films and 1.85 eV for films with predominance of the diamond-like phase) and the dark conductivity by more than 10 orders of magnitude (between 7 and $2 \cdot 10^{-10}\ \Omega^{-1}\ \text{cm}^{-1}$). The dependencies of resistivity and value of optical gap of a–C:H on substrate temperature and the charge power are represented in Fig. 4.7.

Table 4.2 Levels of Structural Modification.

Level	Structural Changes	Method of Treatment	Characterization of Sensitive Properties	Groups of Sensitive Properties	Examples of Sensitive Properties
1	Short range order	Various methods and modes of preparation	All properties	All properties	All properties
2	Medium range order	External factor treatment during preparation or thermal treatments	Properties associated with rearrangement of structural units	Mechanical properties, phase transitions	Viscosity, hardness, Young modulus, photo- contraction of films, temperature and activation energy of crystallization
3	Morphology	Changes in the preparation and treatment modes	Properties dependent on heterogeneities	Electrical, optical	AC conductivity
4	Defect subsystem	Changes in preparation modes, treatments affecting the defect subsystem	Properties dependent on the distribution of localized states density and on the Fermi level position	Electrical, photo- electric	Field dependent conductivity

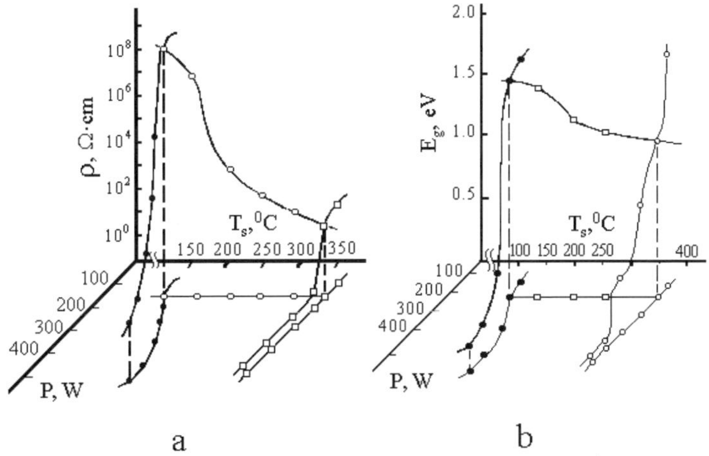

Figure 4.7. Resistivity (a) and optical gap E_g (b) of a–C:H films in relation to their preparation modes (T_s is the substrate temperature, P is the rf discharge power).

Thus the first level of structural modification involves pronounced changes in the short range order of atomic arrangement i. e. changes in the hybridization of electron orbitals of all (or most) atoms constituting the sample. Carbon has long been considered the only (and unique in this respect) element of those considered here that exists in allotropic crystalline modifications of diamond (sp^3 hybridization) and graphite (sp^2 hybridization). The discovery of carbines (sp hybridization) in 1960 (Sladcov *et al.*, 1972) only served as more evidence of the unique properties of carbon. At the same time there have been reports that the short range order may change substantially under certain conditions for other elements as well. For example, cubic α and β modifications of crystalline selenium with atom coordination numbers 4 and 6 respectively were obtained (Andrievsky, Nabitovich and Krinykovich, 1959, Andrievsky and Nabitovich, 1960) in electron beam induced crystallization of thin films. However, these reports failed to attract due attention of the scientific community at that time. The monopoly of carbon on the possibility of existence of forms with different hybridizations of electron orbitals and, consequently, with different atom coordination numbers was radically broken up by series of investigations (Khohlov, Mashin and Khohlov, 1998, Mashin and Khohlov, 1999). New forms of silicon that appeared under certain conditions in films of amorphous silicon and in -Si: were discovered in these studies — silicine with *sp* hybridization of electron orbitals and atom coordination of 2 and form with sp^2 hybridization and atom coordination of 3. However, it should be noted that in these investigations devoted to silicon and in studies of selenium, the new forms were obtained only under certain, rather specific conditions.

Thus, in the considered group of covalent semiconducting materials (periods 2–5, Groups IVA–VIA of the periodic table), structural changes at the short range-order level are observed for elements belonging to Group IVA, periods 2 and 3

(C, Si), and Group VIA, period 4 (Se) (Table 4.3), i.e., for three elements out of the considered nine. We should also note the increase in the first coordination number from 2 to 3 in tellurium melt (at 600°C), with the covalent nature of chemical bonds being preserved (Fig. 4.8, Poltavzev, 1984) and the changes the short range order (bond length) in glassy GeS_2 under pressure (Miyauchi *et al.*, 2001). These elements show no fundamental distinctions in electron shell structure or other parameters from the rest of the considered elements. In view of the above, it seems reasonable

Table 4.3 Short range order level.

Periods	Columns / Coordination number (N_c)								
	IVA		N_c	VA		N_c	VIA		N_c
	6		2						
2		C	3						
	12		4						
	14		2	15			16		2
3		Si	3		P	3		S	
	28		4	31			32		
	32			33			34		2
4		Ge	4		As	3		Se	4
	72,5			75			79		6
				51			52		2
5					Sb	3		Te	3^a
				122			128		

^{a}–melt

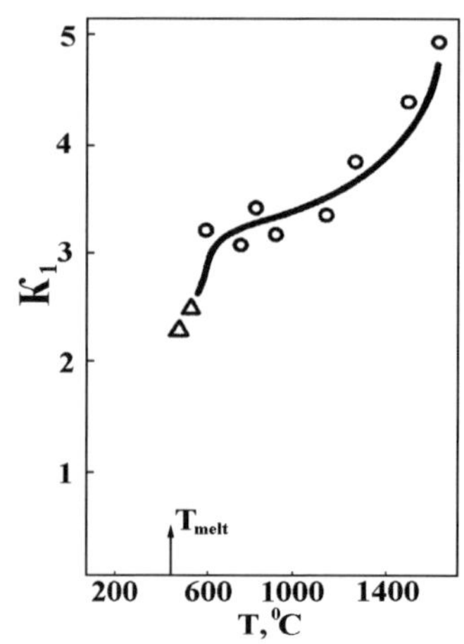

Figure 4.8. Temperature dependence of the liquid tellurium first coordination number (Poltavzev, 1984).

to assume that structural changes at the short range order level are not unique to carbon, being characteristic to all of the considered covalent semiconducting materials. Experimental evidence in favor of this assumption has been obtained for carbon, silicon, and selenium, and obtaining it for the other elements is only matter of time and attention devoted to the problem.

Structural modification at the level of short range order not only determines considerable changes in the properties of a material but also influences upon other groups of structural characteristics, namely: the medium-range order, the morphology, the defects' subsystem. It was shown (Miyauchi *et al.*, 2001) that the changes in tetrahedron structure of glassy GeS_2 caused by high pressure, that is changes of short-range order lead to the change of tetrahedrons' mutual packing, and so to the change of medium-range order. And more significant changes of all the structural characteristics should be expected at the transition from the tetrahedral coordinated network to the layer structure or to the linear polymers.

4.4.2 Structural Modification at the Medium Range Order Level

This level of structural modification implies changes of dihedral angle distribution, degree of molecule polymerization, extent of disorder in alloys, etc. Changing the medium-range order without changing the short-range order mainly affects the macroscopic properties of material (viscosity, micro-hardness, Young modulus, photo-contraction of films–see Table 4.2). At the same time, properties governed by the electronic structure (electronic spectrum) of material change relatively weakly (these depend mainly on the short range order).

In the case of modification at the level of medium range order property control efficiency is determined by flexibility of structural network. The latter decreases with increasing degree of covalent coupling of structural network (average number of covalent bonds per atom $N/2$, where N is average coordination number) and increases with the increasing average number of lone-pair electrons per atom ($[NVE - N]/2$, where NVE is average number of electrons in the outermost shells of the atoms). Thus, flexibility of structural covalent bonded network varies directly as ratio ($[NVE - N]/N$). At the same time the rise of chemical bond ionicity or degree of bond metallization decreases directional character of chemical bonds and increases structural network flexibility. Thus, it is necessary to establish the proportion of ionic and metallic bond components.

Ionicity of chemical bond can be determined by electronegativity difference of atoms as follows:

$$S = 1 - \exp(-\Delta x^2/4) \tag{4.2}$$

where $\Delta x = x_A - x_B$ is electronegativity difference of atoms A and B. In the case of multi-component material, average coefficient of bond ionicity is given by

$$I_C = \sum S_n N_n \tag{4.3}$$

where n is number of various chemical bonds in the material, S_n is the ionicity of bond n and N_n is the fraction of n bonds in the material. Metallicity of the bonds increases with the increase of atomic number in the periodic table. However, it is rather difficult to estimate metallicity in terms of this quantity. As a rule, the degree of bond metallization (M) is taken to be proportional to the average number of electrons per atom. Thus, the covalent coupling of material structural network is characterized by the parameter $N(1 - I_C - M)/2$ and flexibility of structural network varies in proportion to

$$\frac{NVE - N}{N(1 - I_C - M)} \tag{4.4}$$

As a rule, the structural modification at the medium range order level is realized at the temperature range above glass transition temperature. Therefore, property control efficiency depends not only on obtained value of structural changes but also on preservation of these changes during cooling process of material. At an infinitely high cooling rate the number of various structural units and their ratio in the material are preserved the same as in the melt before cooling. However structural relaxation of material can take place during cooling process at the finite cooling rates and the rate of structural relaxation increases with increasing melt temperature. It is necessary to increase cooling rate to neutralize structural relaxation process. The boundary cooling rate needed to avoid structural relaxation depends on temperature as follows (Dembovsky and Chechetkina, 1981, a):

$$V_b = \nu_0(RT^2/\Delta G^*)\exp[-\Delta G^*/(RT)] \tag{4.5}$$

where ν_0 is vibration frequency of molecular groups, R is universal gas constant and ΔG^* is free energy of the initial reorientation of molecular groups (Dembovsky and Chechetkina, 1981, b).

The real cooling rate depends on mass, shape and heat capacity of the sample and on cooling conditions:

$$V_q = T/\tau = (T_0/\tau)\exp(-t/\tau) \tag{4.6}$$

where T_0 is initial melt temperature, t is cooling time and τ is cooling time constant

$$\tau = mc/s\lambda \tag{4.7}$$

where m and s are mass and surface area of cooling sample, c is material specific heat and λ is convective heat transfer factor.

The condition of structural relaxation absence is given by

$$V_q \geq V_b. \tag{4.8}$$

Maximum value of cooling time constant needed to avoid structural relaxation may be obtained from the equality of Eqns (4.5) and (4.6):

$$\tau_{max} = (\Delta G * /\nu_0 RT)\exp[\Delta G * /(RT)]. \tag{4.9}$$

Equation (4.9) allows to formulate demands to preserve structure of the melt in solid sample after quenching.

The value of maximum possible change in the structure, and therefore in the properties of the material, can be written as

$$\Delta\Phi = f\{(NVE - N)/[N(1 - I_C - M)]\}_{Vq \geq Vb} \tag{4.10}$$

where f is unspecified function and $(NVE - N)/[N(1 - I_C - M)]$ is the criterion of structural modification efficiency (CSME) at the medium range order level. The physical meaning of Eqn. (4.10) is expressed by the fact that possibility of obtaining various structural configurations increases with an increase in the degree of freedom of structural units and of the number of long-pair electrons capable of forming unusual structural units in a given material.

The dependence of microhardness changes for a number of chalcogenide glasses as a consequence of changes in the thermal histories of materials as a function of CSME is shown in Fig. 4.9 (Popov, Michalev and Shemetova, 1983). Various thermal histories were obtained by changing the melt temperatures before cooling in the temperature range satisfying the requirements (4.8). Data are provided in percentages per one degree of the melt temperature change.

One can see from Fig. 4.9 that the value of microhardness change increases linearly with increasing CSME:

$$\Delta H = A_H \left[\frac{NVE - N}{N(1 - M - I_C} - B_H \right] \tag{4.11}$$

where A_H is proportionality factor and B_H is the boundary CSME value below which it becomes impossible to change the material property by structural modification at this level. The question arising in connection with this is 'what determines the applicability limit of the structural modification at the medium range order level and where does it lay?'

Qualitatively, it may be stated that this limit is determined by the rigidity of the structural network: for changes in the medium range order to occur, the network

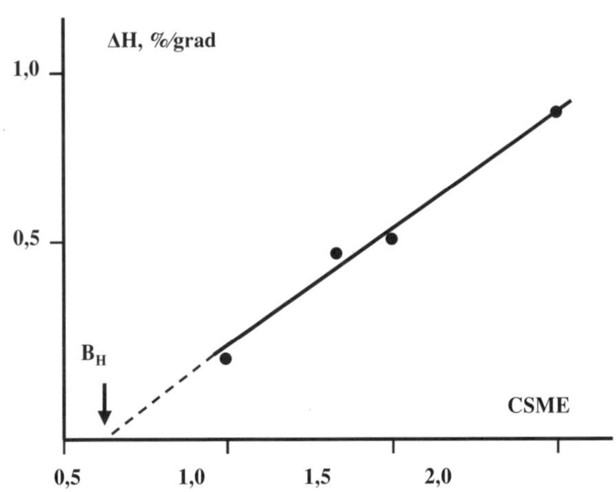

Figure 4.9. Microhardness changes as a function of CSME for chalcogenide glasses with various thermal histories of materials(see text).

must exhibit certain flexibility and flexible bonds are to be present. In view of the above, it seems reasonable to suggest that the limit in question corresponds to the rigidity threshold of the structural network (Philips, 1979a, 1979b, 1980) at which the average number of force constants per atom (N_{const}) becomes equal to the number of degrees of freedom ($N_d = 3$):

$$N_{const} = N_d = 3. \tag{4.12}$$

The number of force constants depends on the type of chemical bonds. In the case of non-directed bonds stretch force constants are present only and $N_{const} = N/2$ (where N is coordination number). Since bond-bending modes are also present in covalent bonded materials, the number of force constants is given by

$$N_{const} = N/2 + [N(N-1)]/2 = N^2/2 \tag{4.13}$$

and the critical coordination number (N_C) corresponding to the rigidity threshold of the covalent bonded structural network is defined by

$$N_C^2/2 = 3 \text{ or } N_C = 2.4.$$

The values of average coordination number for materials of systems $A_x^{IV}(A_x^V)B_{1-x}^{VI}$ are shown in Table 4.4. The rigidity threshold ($N_C = 2.4$) is shown

Table 4.4 Medium range order and morphology levels.

Columns		A^{IV}_x (A^V_x) B^{VI}_{1-x} systems Value of X							Column VI A
IV A	V A	0,60	0,50	0,40	0,33	0,25	0,20	0,11	
Ge **4** ***4***					**$GeSe_2$** **2,66** ***2,40***	**$GeSe_3$** **2,50** ***2,30***	**$GeSe_4$** **2,40** ***2,24***	**$GeSe_8$** **2,22** ***2,12***	**Se** **2** ***2***
Ge **4** ***4***			**GeS** **3,00** ***2,73***	**Ge_2S_3** **2,80** ***2,49***	**GeS_2** **2,66** ***2,30***	**GeS_3** **2,50** ***2,23***	**GeS_4** **2,40** ***2,19***	**GeS_8** **2,22** ***2,10***	**S** **2** ***2***
	As **3** ***3***	**As_3Se_2** **2,60** ***2,53***	**AsSe** **2,50** ***2,42***	**As_2Se_3** **2,40** ***2,31***	**$AsSe_2$** **2,33** ***2,25***	**$AsSe_3$** **2,25** ***2,19***	**$AsSe_4$** **2,20** ***2,16***	**$AsSe_8$** **2,11** ***2,09***	**Se** **2** ***2***
	As **3** ***3***	**As_3S_2** **2,60** ***2,49***	**AsS** **2,50** ***2,38***	**As_2S_3** **2,40** ***2,26***	**AsS_2** **2,33** ***2,22***	**AsS_3** **2,25** ***2,17***	**AsS_4** **2,20** ***2,14***	**AsS_8** **2,11** ***2,08***	**S** **2** ***2***

▬ ▬ ▬ - rigidity threshold in the case of purely covalent bonds;
▬▬▬▬ - rigidity threshold in the case of bond ionicity is taken into account;
2,25 - average coordination number, N;
2,25 *(italics)* - $N(1 - I_c)$, where I_c – mean coefficient of bond ionicity.

by the dashed line. The materials with flexible structural network are on the right hand side of that line. Structural modification at the medium range order level is possible in these materials. To the left of the dashed line, there are materials for which structural modification at this level is impossible because of their rigid structural network. However, experimental studies of these systems (Yun *et al.*, 1989; Aitken, 2001) have shown that the rigidity threshold actually lies at higher average coordination numbers. It was demonstrated (Popov, 1994) that the reason for the discrepancy between the calculated and experimental values of the rigidity threshold comes from the ionic component and metallization of chemical bonds being neglected. The rigidity threshold obtained for the above-mentioned systems taking into account ionicity of chemical bonds is shown in the table by the solid line. In accordance with this suggestion, the line is the applicability limit of structural modification at the medium range order level.

It is necessary to note that a change in material's thermal history is not the only way of modifying the structure at the middle range order level. Combined effect of thermal and other external factors (electric and electromagnetic fields, ultrasonic vibrations and others) increases the structural modification efficiency at the middle range order level.

4.4.3 Structural Modification at the Morphology Level

The necessary condition for obtaining any solid non-crystalline material is the thermodynamically non-equilibrium process of its synthesis. In conformity with the basic concepts of the theory of self-organization (synergetics), the non-equilibrium conditions of material formation result in the appearance of heterogeneities. Thus, general considerations suggest presence of heterogeneities in all of the non-crystalline semiconductors in question.

The presence of macroheterogeneities (columns, globules, cones etc.) on a scale of tens nanometers and more in the non-crystalline material films is considered as film morphology. Certain morphology (columns structure) was observed in hydrogenated amorphous silicon films by Knights and Lujan in 1979. It was demonstrated later that certain morphology is a characteristic feature of all amorphous tetrahedral semiconductors. At the same time a 'structureless' smooth surface is commonly observed in vitreous materials. Considering that the presence of heterogeneity is in the nature of non-crystalline solids themselves, one can conclude that in glassy materials there are microheterogeneities. It points to the manifestation of not the morphology, but medium range order (Popov, 2002). However, such a conclusion implies the answer to a number of questions, namely: why are the two classes of non-crystalline semiconductors characterized by various kinds of heterogeneity? What factors cause the separation between them and where does such boundary passes? To answer these questions let's return to the medium range order of atomic arrangement.

Medium range order is linked with correlation between atom positions in the range of up to one or a few nanometers. Models of medium range order (see Sec. 2.2) can be divided into two groups. In the first group the ordered regions determining medium range order are the part of surrounding matrix (or they are linked with each other) without breaking the continuous network. In the second group between the ordered regions there are limits at which structural network breaks. The possibility for the first or the second situation to be realized in a specific non-crystalline material depends upon the flexibility of its structural network. The limit in question corresponds to the rigidity threshold of the structural network at which the average number of force constants per atom becomes equal to the number of degrees of freedom (see Sec. 4.4.2). In the first case the network must exhibit certain flexibility. Materials in which the formation of the ordered regions determining the medium range order (microheterogeneities) is made possible without breaking the continuous covalent network are placed on the right hand side of the solid line in Table 4.4. As stated above structural modification at the level of medium range order is possible in these materials.

The materials with rigid structural network are on the left hand side of the solid line. Formation of the ordered regions in these materials leads to the break of continuous covalent network and to formation of limits between microheterogeneities. While synthesizing such materials, some microheterogeneities can reach the size of macroheterogeneities forming the specific morphology. Hence in the considered materials structural modification is made possible at the level of morphology. Morphology changes affect the properties sensitive to macroheterogeneities (see Table 4.2).

Thus, the limit between medium range order and morphology levels of structural modification is the rigidity threshold of structural network. The morphology level of structural modification can be applied only to the materials with structural network rigidity higher than the above mentioned rigidity threshold.

4.4.4 Structural Modification at the Defect Subsystem Level

This level of structural modification is associated with changes in the defect subsystem under the influence of either sample fabrication conditions or various external factors and is manifested by changes in the spectrum of localized states in the gap, which in turn lead to a shift in the Fermi level position and modify the properties related to the electron subsystem.

Selection of the influence factor is important in this case because it has to interact with defects but not with structural network on the whole. It has been shown (Popov, 1996) that treating hydrogenated amorphous silicon films with ultraviolet radiation leads to significant change in the spectrum of localized states in the material, resulting from a change in the relative content of $Si-H$, $Si-H_2$ and $Si-H_3$ complexes. Dependences of infrared absorption peak areas corresponding to various complexes $Si-H_n$ in a–Si:H on ultraviolet radiation dose are shown in Fig. 4.10. Area of peak at 640 cm^{-1} corresponding to $Si-H_3$ complexes decreases while area of peak at 610 cm^{-1} corresponding to $Si-H$ complexes increases with increasing ultraviolet radiation dose. Area of peak at 590 cm^{-1} corresponding to

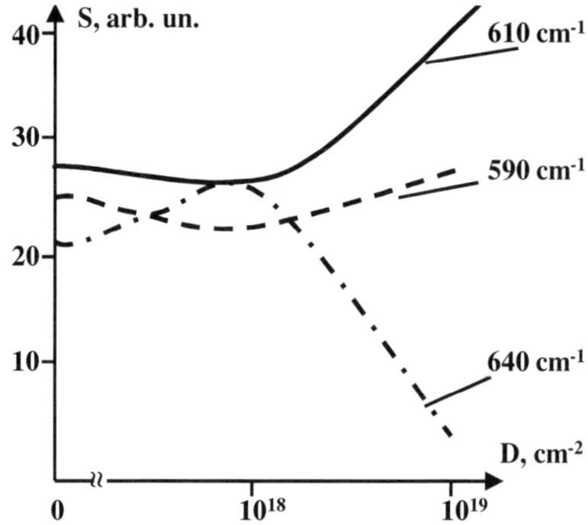

Figure 4.10. Dependences of infrared absorption peak areas corresponding to various complexes Si–Hn in a–Si:H on ultraviolet.

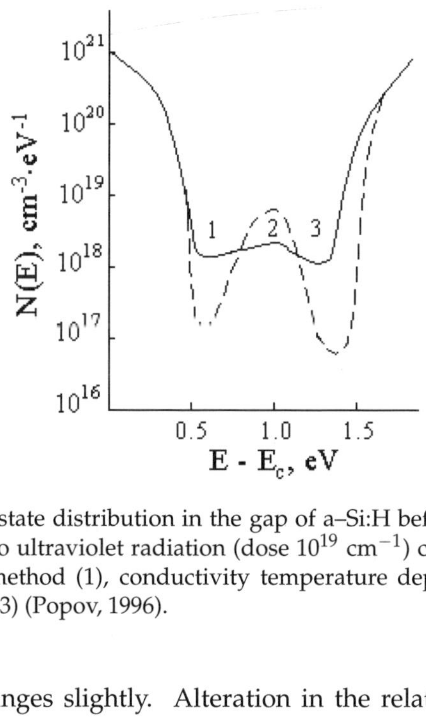

Figure 4.11. Localized state distribution in the gap of a–Si:H before (solid line) and after (dashed line) exposure to ultraviolet radiation (dose 10^{19} cm^{-1}) calculated from results of constant photocurrent method (1), conductivity temperature dependence (2) and space-charge-limited currents (3) (Popov, 1996).

$Si-H_2$ complexes changes slightly. Alteration in the relative content of $Si-H_n$ complexes leads to changes of localized state distribution in the gap (Fig. 4.11). One can see that the length of valence and conduction band tails decreases and the density of localized states in the middle of the gap increases under the influence of ultraviolet radiation. In turn, this leads to enhanced stability of a–Si:H films and to improvement of thin film transistor parameters.

The effect of amorphous silicon pseudo-doping that varied with film preparation conditions has been observed by Golicova (1991). Dembovskii, Chechetkina

Table 4.5 Defect subsystem level.

Columns		Systems		Column
IV A	V A	$A_2^V B_3^{VI}$	$A^{VI} B^{VI}$	VI A
C				
Si	P			
Ge	As	As_2S_3	S - Se	S
		As_2Se_3	Se -Te	Se
		As_2Te_3		Te

and Kozukhin (1985) observed the effect of weak electric and magnetic fields on quasimolecular defects (see Chapter 3) and properties of vitreous selenium, arsenic triselenide and materials of the selenium– tellurium system. Dembovsky, Zyubin and Grigor'ev (1998) extended these results to sulfur and arsenic trisulfide. Thus, the defect subsystem level of structural modification is observed experimentally both in a tetrahedral material with rigid covalent structural network (a–Si:H) and in vitreous materials of group VI and V–VI chalcogenide glasses (Table 4.5). The above suggests that structural modification at this level is inherent, as is the structural modification at the short range order level, in all of the considered non-crystalline semiconducting materials.

It can thus be concluded that there exist four levels of structural modification of the properties of non-crystalline semiconductors that differ in the resulting structural changes. Control over properties of non-crystalline semiconductors by changing the structure at the short range order level is, in principle, possible for all of the considered materials. Control over properties by treating the defect subsystem is also applicable to all these materials. Control over properties by changing the medium range order is possible for vitreous semiconductors that have high values of the criterion of structural modification efficiency, with the applicability limit of this level corresponding to the rigidity threshold of the structural network, calculated with consideration of ionicity and metallization of chemical bonds. On the other side of this boundary (materials with rigid structural network) lies the area of applicability of structural modification, involving changes in the morphology of materials.

4.4.5 Correlation Between Structural Modification and Stability of Material Properties and Device Parameters

Structural modification (changes of structure) can take place in the production process as well as in the process of affecting material or devices on the basis of different factors (Table 4.2). Moreover, atomic structure and consequently properties of non-crystalline semiconductors can change in the process of using such devices because their operation is subjected as a rule to the influence on the material by electrical or electro-magnetic fields, high temperature and so on. Two conclusions follow from the above and these must be taken into consideration when designing devices on the basis of the observed materials.

The first conclusion consists of the possibility to improve parameters of such devices by purposeful effect on the structure of material during or after their production. The present approach has been put into practice during the creation of electrophotographic photoreceptors on the basis of glassy selenium (thermal treatment and electromagnetic radiation), integrated circuit of reprogramming memory matrix on the basis of chalcogenide glassy semiconductors (γ-emission), photo-thermoplastic optical information memory units on the basis of chalcogenide glasses (thermal treatment), thin film transistors on the basis of hydrogenated amorphous silicon (ultraviolet emission) (Popov, 2004).

The second conclusion is that the possibility of changes in non-crystalline semiconductor structure during the device operation leads to statistical variations of device characteristics that cannot be avoided or eliminated. The value of these variations will grow in line with the level of structural changes possible in non-crystalline semiconductors used at the given levels of external factors. Dependence of statistical variations of threshold voltage in various chalcogenide glassy semiconductors switches on the value of criterion of structural modification efficiency is shown in Fig. 4.12 (Popov, Michalev and Shemetova, 1983). As seen in this figure, experimental data of statistical variations of threshold voltage maps well to power law dependence on the value of criterion efficiency of structural modification.

Thus in many cases structural modification of properties of non-crystalline semiconductors is an effective method of device parameter control. But it is also necessary to take into consideration that possibility of structural changes of non-crystalline semiconductors in the period of device operation defines the value of unavoidable variation of device parameters.

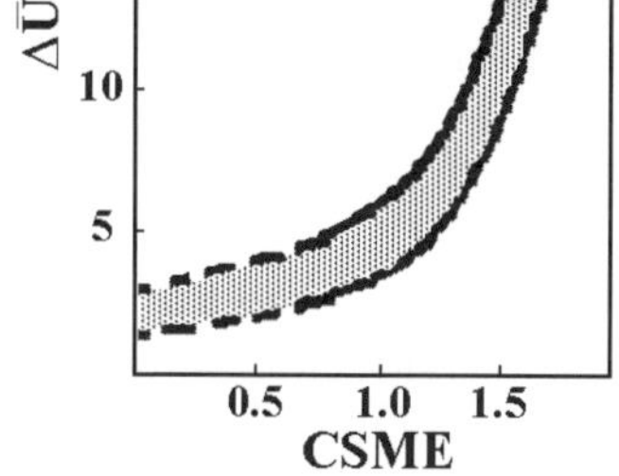

Figure 4.12. Statistical variations of threshold voltage in various chalcogenide glasses as a function of the criterion for structural modification efficiency.

Chapter Five

Preparation Methods of Disordered Semiconductor Films

Use of disordered semiconductors in a device is usually limited to thin films. Such films must have predefined properties and the preparation methods must guarantee the reproducibility of these properties. Not surprisingly, different methods are used for the preparation of different disordered semiconductors. For example, thin films of chalcogenide glassy semiconductors are manufactured by thermal vacuum evaporation and radio-frequency sputtering methods, while methods of radio-frequency or DC glow discharge gas decomposition, thermal or catalytic chemical vapor deposition and reactive sputtering are usually used for preparation of hydrogenated amorphous silicon films.

5.1 TECHNOLOGICAL DISTINCTIONS OF CHALCOGENIDE GLASSY FILM PREPARATION

The most common method of thin film preparation is thermal vacuum evaporation. In the case of chalcogenide glassy films the distinctions in preparation consists of evaporation methods and evaporator designs. The simplest construction is an open evaporator with direct resistive heating. Boat-type evaporator made of high-melting-temperature metal (tantalum or tungsten) with an ingot of material to be evaporated is heated by electric current in a chamber in vacuum below 10^{-5} Torr. The ingot of material melts under action of Joule heat, turns into a vapor and deposits on substrate. Film thickness is controlled by time of deposition with the use of a shutter in front of substrate. This type of evaporator is used in the case of (a) elementary materials which have relatively low melting temperature (for example selenium) and (b) in the case of stable chemical compositions (for example stoichiometric compositions As_2S_3 and As_2Se_3,).

The majority of chalcogenide glassy semiconductors is two- or multi-component materials. These materials are inclined to thermal dissociation.

Disordered Semiconductors: Physics and Applications by Anatoly Popov

www.panstanford.com
978-981-4241-76-2

Moreover, components of these materials have different values of vapor pressure and therefore different evaporation rate. As a result using open evaporator with direct resistive heating leads to quantitative change of chemical composition along the thickness of the film. (Instead of initial *AB* composition in the evaporator the film may be "a sandwich" of *A* and *B* components). Yet, in some cases, it is possible to prepare film of two-component material by using this type of evaporator. Various chemical elements have different slope of curves of saturated vapor pressure on temperature dependencies. If these curves cross at acceptable temperature value then evaporation rates of components are equal at this temperature. However, technological process is intrinsically unstable in this case. Furthermore, questions on the meaning of "acceptable" temperature and determination of the evaporation temperature of chalcogenide film preparation process remain.

The evaporation temperature defines vapor pressure and thereby film growth rate. Firstly, film growth rate should not be too slow or too fast. Secondly, thermal dissociation probability of multi-component material depends on evaporation temperature as well. Considering the above, the decrease of evaporation temperature seems to be advisable. On the other hand, the kinetic energy of vapor particles increases with increasing evaporation temperature. This results in improved adhesion of film which is important from the point of view device reliability. The listed considerations conflict with each other even for simple compositions. And very often it is impossible to find acceptable compromise for multi-component chalcogenide glassy film preparation. Therefore, special evaporator designs and evaporation methods are employed.

Methods of flash evaporation are based on very rapid (momentary) heating and evaporation of material before the process of thermal dissociation takes place. There are various ways to realize this idea. In one case (the large current pulse evaporation) capacitor *C* is charged by voltage source *U* and closed to evaporator (Fig. 5.1(a)). Large current pulse heats evaporator instantly up to high temperature. However in the case of open-boat evaporator rapid heating leads to splash out of material due to "vapor cushion" formation between evaporator walls and the evaporated material. To prevent this two covers with displaced outlets cover evaporator (Fig. 5.1(b)). This method can only produce film with limited thickness, which is an obvious shortcoming.

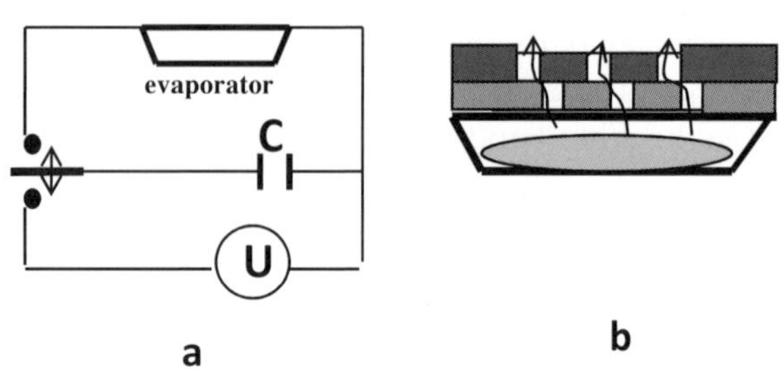

Figure 5.1. Large current pulse evaporation diagram (a) and evaporator design (b).

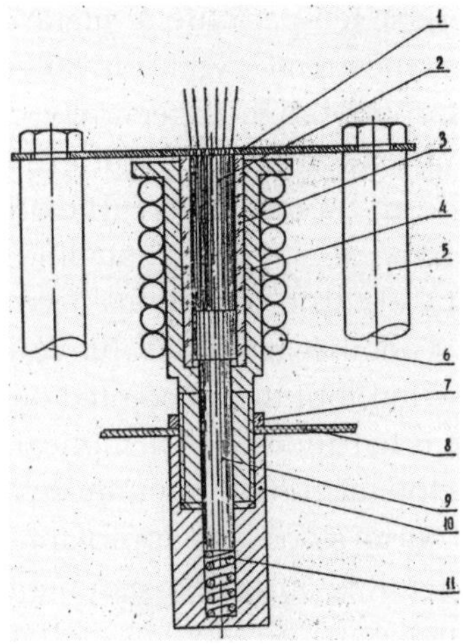

Figure 5.2. Evaporator with continuous feeding of evaporated material for chalcogenide films evaporation (1 is heated part of evaporator, 2 is evaporated material, 3 is quartz tube, 4 is upper part of body, 5 is current lead, 6 is water cooling system, 7 is nut, 8 is basis, 9 is piston, 10 is lower part of body and 11 is spring).

Evaporator with vibrating bin is free from this shortcoming. In this case vibrating bin feeds small portions of material on evaporator surface heated to high temperature. These portions evaporate instantly on contact with the heated surface. There is no limitation of film thickness. However the determination of temperature conditions is a rather complicated task in this case. Underheating of evaporator leads to decomposition of the material, whereas overheat leads to the splash effect.

A solution to these constraints in the case of chalcogenide glassy films is provided by an evaporator with continuous feeding of material to be evaporated (Fig. 5.2) (Popov, 2008). The heated part of evaporator (1) is tantalum or tungsten band with holes. Piston (9) presses evaporated material (2) to the heated part of the evaporator. Vapor passes through the holes in evaporator band. Evaporator body (4) has water cooling system (6) and so only upper part of evaporated material reaches evaporation temperature. Under these conditions vapor composition does not depend on evaporation time or on the amount of the evaporated material. As a result chemical composition of the film remains constant in the process of preparation.

The other group of flash evaporation methods is laser induced evaporation and electron beam evaporation. The essence of the former is the evaporation of material by short high intensity laser pulses.

A small amount of material is heated up to 10^6 °C during laser pulse time of $10^{-7} - 10^{-8}$ sec. This method is characterized by

- identical chemical composition of the vapor and the material being evaporated,
- ability to obtain high film growth rates (up to 6 μm/min) with exact control of film thickness by number of laser pulses,

- possibility to obtain super thin films with a single pulse or to produce super lattice,
- conversion of evaporated material into plasma due to interaction with electromagnetic radiation and high temperature and deposition of film from plasma,
- absence of heating of vacuum chamber parts and minimal film pollution.

Laser induced evaporation method retains the shortcoming of the splash effect. Characteristics of electron beam evaporation method are similar to laser induced evaporation method. However, technical realization is more complicated in this case because electron gun has to be inside of the vacuum chamber.

One of the ways to solve the contradiction between the necessity to decrease evaporation temperature for minimization of multi-component material dissociation process and the necessity to increase it for adhesion improvement is evaporation from quasi-closed volume method. The concept behind this method is illustrated in Fig. 5.3.

The evaporator consists of lower part 1 and cover 2 with capillary hole 3 on the top. Bottom of the evaporator is heated to temperature T_1 which is enough for material melting only. Vapor goes out from evaporator through capillary that's why vapor pressure increases in the evaporator space over the melt. This part of evaporator 4 is heated to temperature $T_2 > T_1$. It leads to the rise in vapor particles' kinetic energy and to increase of film adhesion.

The methods described above have the same algorithm. At first, bulk chalcogenide glass of required chemical composition is synthesized. After that various methods are used to prevent dissociation of synthesized material during evaporation process. On the other hand if these materials are inclined to dissociation it may be better to exclude the process of synthesis at all. This idea was realized in the method of co-evaporation from different sources. Let us consider preparation of chalcogenide film with A_xB_y composition by this method. Components A and B are loaded in two evaporators. Evaporator temperatures are selected from the condition that ratio of component evaporation rates (V) is equal to required ratio

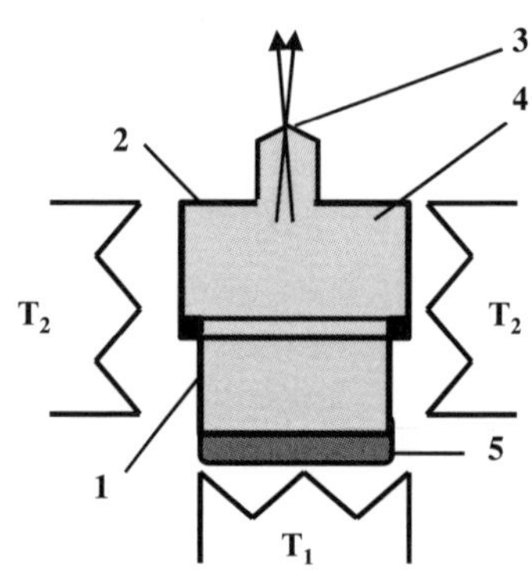

Figure 5.3. Schematic illustration of evaporator with quasi-closed volume (1 is evaporator body, 2 is vacuum-tight cover, 3 is capillary, 4 is vapor, 5 is melt and T_1, T_2 are heaters).

of component concentration in the film:

$$V_A(T_A)/[V_B(T_B)] = x/y \tag{5.1}$$

The condensed film composition is equal to $A_xB_y$under this condition. It is necessary to note that it is possible to prepare films with a determined chemical composition profile along the thickness of the film by temperature changes of one or both evaporators during the evaporation process.

On the whole thermal vacuum evaporation methods have a number of shortcomings:

- changes of chemical composition in the process of multi-component chalcogenide film preparation,
- rather poor film adhesion to substrate,
- difficulties in preparation of large area films with even thickness because vapor source has point or linear form,
- sluggishness of process in the case of resistive heated evaporators.

It is possible to solve all these problems but it leads to significantly more complex technological equipment. To a certain degree the sputtering methods are free from these shortcomings.

Sputtering methods include the following modifications:

- direct current (cathode) sputtering,
- magnetron sputtering,
- radio-frequency sputtering,
- reactive sputtering.

Schematic illustration of diode-type direct current sputtering is shown in Fig. 5.4. Cathode (1) and anode (2) are located in vacuum chamber (3). Target of sputtered material (4) is placed on cathode and substrates (5) are placed on

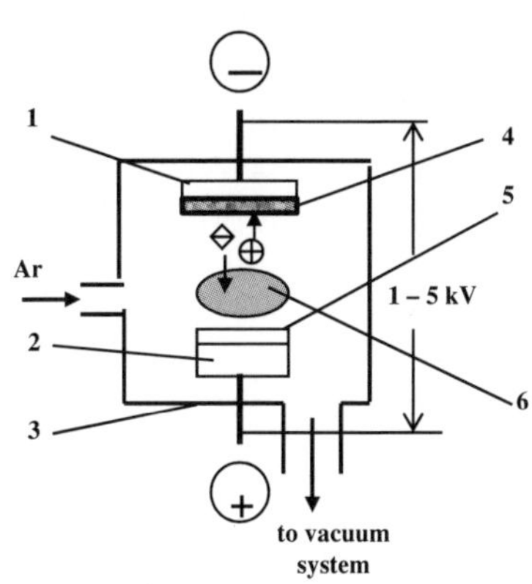

Figure 5.4. Schematic illustration of diode-type direct current sputtering reactor (1 is cathode, 2 is anode, 3 is vacuum chamber, 4 is target of sputtered materials, 5 are substrates and 6 is glow discharge).

anode. After air evacuation argon leak-in vacuum chamber up to pressure 10^{-2} Torr approximately. High voltage applied to electrodes leads to electron emission from cathode. Electrons ionize argon atoms and glow discharge (6) is initiated between electrodes. Positive argon ions are accelerated in electrical field and bombard cathode-target. As a result target surface sputters and material particles deposit on the substrates.

Sputtering method has the following features.

- Because energy of bombarded ions (keV) is considerably higher than the energy of chemical bonds (eV) various materials have close sputtering rates. In the case of diode-type system sputtering rates are 100 – 300 Å/min.
- The same reason leads to high energy of sputtered particles and high film adhesion to substrate.
- Relatively low sputtering temperature prevents dissociation of material and ensures low chemical composition changes of film in comparison with target of sputtered material.
- It is easier to solve the problem of preparation of large area films with even thickness because source has flat form.
- Addition of chemically active gases (oxygen, nitrogen etc.) to argon in the vacuum chamber during elementary material sputtering allows to prepare compound film with these gases (oxides, nitrides, carbides etc.). This method is called the reactive sputtering method.

Diode-type direct current sputtering reactor has relatively slow rate of film growth. Sputtering rate is determined by concentration of bombarded target ions. The latter depends on concentration of cathode emitted electrons and on gas pressure in reactor. Because electron emission from cold cathode is small enough it is necessary to grow gas pressure up to 10^{-2}–10^{-1} Torr. On the other hand increasing gas pressure leads to a rise in sputtered particle scattering by gas atoms and to a decrease of film growth. Thus it is not possible to increase film growth rate within the limits of diode-type direct current sputtering.

Triode-type direct current sputtering reactor has been developed to solve this problem. The main difference consists in division of cathode and target functions. In other words the third electrode — thermo-emission cathode is added to chamber. It increases electron and consequently argon ion concentrations and allows to decrease gas pressure to approximately 10^{-3} Torr. Moreover, it allows to bring substrates nearer to target and to increase film growth rate as well. As a result film growth rate rises up to 1000 Å/min.

However it is possible to increase argon ion concentration without increasing electron concentration by extending electron tracks from cathode to anode and appropriately increasing the number of ionization acts. It is known that electrons spiral in axial magnetic field. Thus the superposition of electrical and magnetic fields leads to an increase of gas ionization efficiency and film growth rate.

Application of cross-wise electrical and magnetic fields in reactor is the basis of magnetron sputtering.

Direct current sputtering is an effective method of film preparation in the case of materials with high electrical conductivity. However it does not give possibility to spurt dielectrics and high resistivity semiconductors. The reason is that the ions bombarding the target accumulate on the target's surface and form positive charge that compensates the applied electrical field. It leads to close sputtering process. Therefore radio-frequency sputtering is used to prepare films of materials with high resistivity. High frequency voltage (10–13 MHz) is applied to target in this case. Thus target is bombarded by positive charge gas ions during the time of half — period and is bombarded by electrons which neutralize ion positive charge during the time of next half-period. In fact, target is bombarded by positive charge gas ions for the duration of time more than half-period as ion mobility is less than electron mobility and the effect of target self-bias takes place. Radio-frequency sputtering allows preparation of high resistivity multi-component chalcogenide glassy films, hydrogenated amorphous silicon films, silicon dioxide and silicon nitrides films and etc.

The methods described above ensure the preparation of films with a given constant chemical composition and with a given profile of chemical composition along the thickness of film. One could think that the problem of films with required properties preparation is solved. But such conclusion is premature in the case of disordered semiconductor films. The absence of long range order in atomic arrangement leads to various atomic structures at constant chemical composition (see Chapter 4) and to different film properties. Therefore properties of disordered material films depend strongly on conditions of film preparation even in the case of invariable chemical composition. The main factors that determine chemical composition, atomic structure and consequently properties of disordered material films are shown in Fig. 5.5.

Film chemical composition (red italics in the figure) is determined by vapor chemical composition. The latter depends on the following factors:

- chemical composition of evaporated material,
- evaporator temperature,
- evaporator design,
- substrate temperature,
- gas pressure and composition in vacuum chamber.

Film atomic structure (blue underlined text in figure) depends to a large extent on the mobility of condensed particles on the substrate. The latter is determined by:

- the substrate temperature,
- the kinetic energy of condensed particles which in turn depends on evaporator temperature and design,

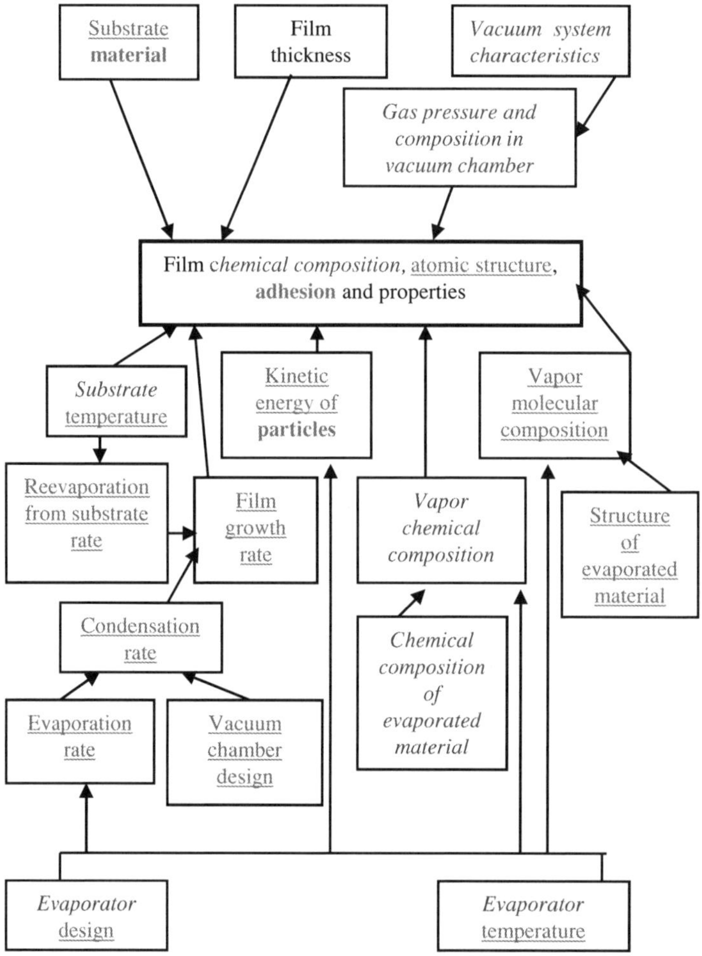

Figure 5.5. Determined film characteristic factors. For color reference, go to page 191.

- the film growth rate which depends on rate of particle condensation and, therefore, on vacuum chamber geometrical characteristics and evaporation rate. The latter depends on evaporator temperature and design as well. Besides that, film growth rate depends on substrate temperature, because re-evaporation from substrate is possible at high temperatures,
- the vapor molecular composition which is determined by evaporator temperature and design and by structure of evaporated material,
- the substrate material.

Film adhesion (green bold text in figure) is determined by kinetic energy of condensed particles, material and surface conditions of substrate. These properties also depend on film thickness.

It is necessary to take into account all of the factors mentioned above when considering device manufacture. Let us consider a solution for these problems in the case of selenium electrophotographic (xerographic) drums (see Chapter 6) production. In this case duralumin drum substrates are used. Dram height has to be no less than the width of duplicated original (300, 420 and 660 mm as a rule) and dram diameters lie within the limits of 25–250 mm. Because an elementary material (selenium) is used in this case we can exclude the film's chemical composition changes.

It is known from experiments that selenium evaporation on cold (room temperature) substrate leads to formation of completely disordered films without any crystalline inclusions. However such films have low value of crystallization activation energy and spontaneous crystallization takes place even at room temperature. On the contrary selenium films prepared at the substrate temperature of 70–80°C have crystalline inclusions but non-crystalline part of film has high value of crystallization activation energy and remains in non-crystalline state at higher temperature. The reason lies in the different degree of molecule polymerization. Mobility of selenium molecules on substrate at room temperature is low. Therefore polymerization of separate molecular fragments into long chain molecule as well as packing of fragments in crystalline lattice is hardly probable. It leads to formation of disordered film consisting of short chain molecules with low activation energy of crystallization. Molecule mobility on substrate increases with substrate temperature. It makes both polymerization process and packing in crystalline lattice easier. As a result, crystalline inclusions arise but non-crystalline part consists of long chain molecules (see Chapter 2) and rearrangement of these molecules into crystalline lattice requires much more energy.

Taking the above into consideration, the following requirements for the selenium drum preparation can be formulated:

- The substrate temperature has to be as high as possible without reaching crystallization temperature. This requires fine control and adjustment of substrate temperature in the vacuum chamber.
- The evaporation temperature has to be low to ensure sufficiently large molecular fragments in vapor phase.
- The kinetic energy of molecular fragments in vapor phase and therefore evaporation temperature have to be high enough to ensure high film adhesion to substrate.

One of the possible variants to solve these contradictory requirements is shown in Fig. 5.6 (Popov, 2008). Vacuum chamber (1) has cylindrical form. Ringed evaporators (2) are situated on the inside surface of chamber wall and are heated by heaters (3). Drum substrate (4) is placed axially in the centre of vacuum chamber. Selenium is charged into evaporators and thermally treated for transformation into crystalline trigonal phase which consists of long chain molecules (see Chapter 2). Selenium sublimation from solid state at the temperature 210–215°C is carried out (trigonal selenium melt temperature is 218°C). Substrate heating is fulfilled by thermal radiation of evaporators. Therefore substrate temperature increases with evaporation time and reaches 80–85°C at the end of film formation

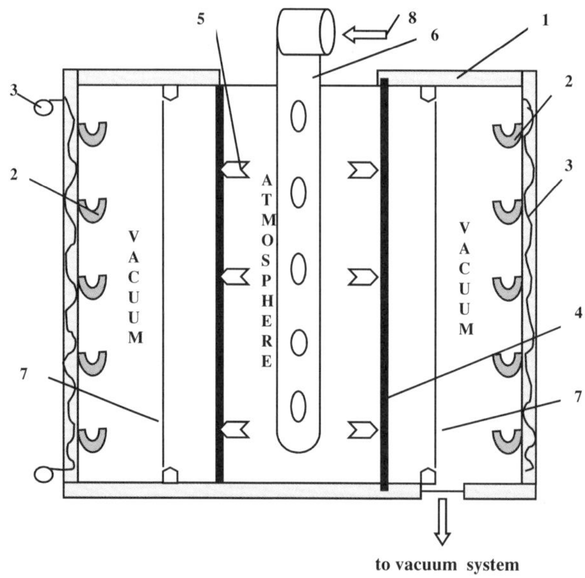

Figure 5.6. Schmetic illustration of xerographic drum preparation system.

process. Sharp substrate cooling is necessary to prevent from film crystallization process in this situation. However it is not easy to realize sharp cooling in the vacuum. That is why drum substrate performs the function of vacuum chamber wall. The insides of drum substrate are vented to atmosphere. It allows accurate substrate temperature monitoring by thermocouples (5) and sharp cooling of substrate by feeding of water-air mixture (8) to insides of dram substrate through pipe-line (6). The last but not least problem that needs to be addressed is the problem of film adhesion. Due to low evaporator temperature (210–215 °C), the kinetic energy of vapor particles is not high enough to guarantee good adhesion. It leads to insufficient adhesion of film to substrate. To eliminate this problem additional string-like heaters (7) are set in vacuum chamber. These heaters are switched on at the beginning of evaporation process for the short time only and heated up to 2000°C. As a result kinetic energy of vapor particles at the initial stage of film formation increases. It leads to a rise of film adhesion to substrate. Electro-photographic drum manufacturing plant and selenium drums are shown in Fig. 5.7.

5.2 PREPARATION OF HYDROGENATED AMORPHOUS SILICON FILMS BY GLOW DISCHARGE DECOMPOSITION METHOD

Glow discharge represents weakly ionized plasma. Concentration of electrons and positive ions is approximately 10^{10} cm^{-3}. This is 10^4 times less than gas molecule concentration at pressure 10^{-1} Torr. Electron energy varies from 1 to 10 eV. This is comparable with energy of chemical bonds. Owing to these characteristics there

b

a

Figure 5.7. Electro-photographic drums manufacturing plant (a) and selenium drums (b). For color reference, go to page 192.

is a wide set of different chemical units in glow discharge plasma: atoms, ions and various free radicals. These units have heightened chemical activity. That is why chemical reactions in glow discharge take place at lower temperatures in comparison with the normal conditions.

The method of radio-frequency glow discharge (13.56 MHz) in silane (SiH_4) or in silane-hydrogen mixture ($SiH_4 + H_2$) is used for hydrogenated amorphous silicon film preparation. There are two types of reactors that differ depending on the way of radio-frequency power input to plasma: reactor with capacitive coupling and reactor with inductive coupling. Reactor with inductive coupling (Fig. 5.8(a)) has been developed first. Silane or gas mixture feeds to fused quartz chamber (1) 5–10 cm in diameter. Vacuum pumping-out is used to create gas flow in chamber. As a rule gas pressure in chamber is equal to 0.1–1.0 Torr and gas flow rate lies within the limits of 0.1–10 cm^3/min. Silane is a fire hazard gas. It spontaneously ignites in the air at concentration of 0.4%. Therefore gas reclamation facility (6) has to be mounted before vacuum system. Substrate (3) is placed on pedestal (4) with heater (5). Gas-discharge plasma is activated by radio-frequency inductor (2). Silane molecules decompose in plasma and silicon atoms precipitate on substrate.

Hydrogenated amorphous silicon film properties depend strongly on plasma parameters, substrate temperature, gas flow rate and geometrical characteristics of the chamber. It has been shown by experiment that substrate temperature has to be within the limits of 250–350 °C and radio-frequency power has to minimize to the value which is necessary to maintain glow discharge to obtain high quality hydrogenated amorphous silicon films (films with low concentration of localized states in the gap). At lower temperatures mobility of atoms on substrate surface

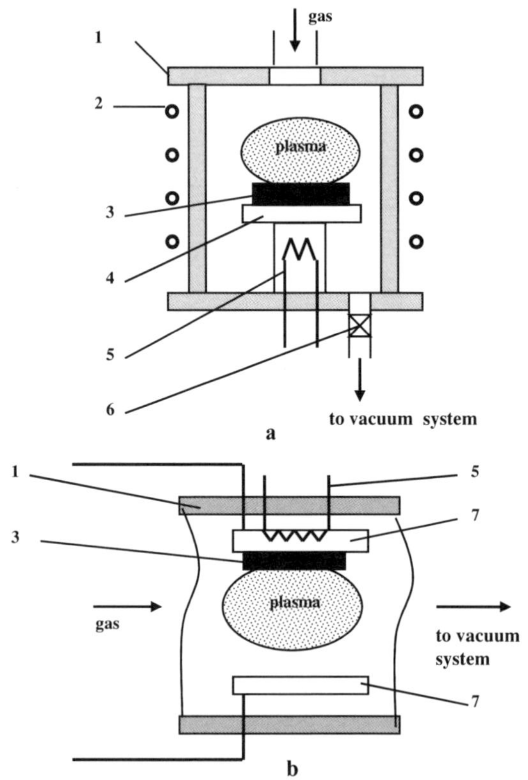

Figure 5.8. Schematic illustration of reactor with inductive coupling (a) and reactor with capacitive coupling (b): 1 is chamber, 2 is radio-frequency inductor, 3 is substrate, 4 is pedestal, 5 is heater, 6 is gas reclamation facility and 7 are electrodes.

decreases. This leads to increased defect concentration in the film. The increase of radio-frequency power leads to the same effect due to higher atom deposition rate. At higher temperatures hydrogen effusion from a–Si:H film takes place. It gives rise to higher concentration of localized states in the gap. However film growth rate is rather low under these conditions (100–200 Å/min). The main shortcomings of reactor with inductive coupling are non-uniformity of properties across film area and impossibility of large area films preparation because increasing the reactor's diameter requires a sharp increase in radio-frequency generator power. That is why the other type of reactor is used in industrial production.

In the case of reactor with capacitive coupling (Fig. 5 8(b)) radio-frequency power is applied to parallel flat electrodes (7) inside the chamber. One can see that it allows to increase electrode areas and therefore to prepare large area uniform films. Parameters of deposition process (gas pressure, flow rate, substrate temperature and radio-frequency power) lay within the same limits as in the case of reactor with inductive coupling.

Devices based on hydrogenated amorphous silicon as a rule consist of different layers with various type and value of conductivity. That is why multi-chamber

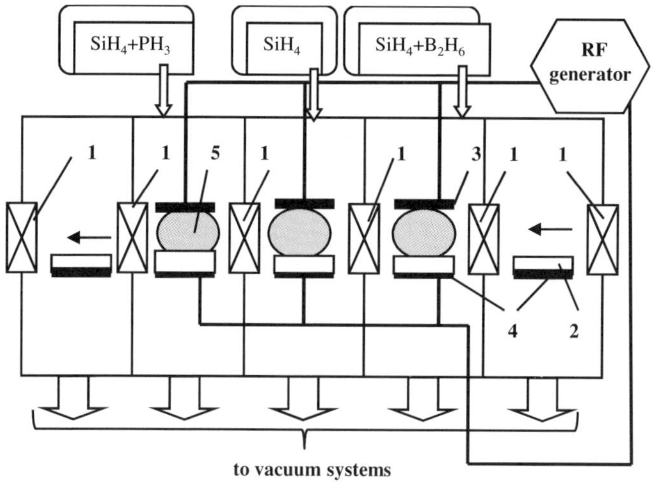

Figure 5.9. Schematic illustration of multi-chamber reactor with semi-continuous process of *p-i-n a* Si:H solar cells production (1 is vacuum seal, 2 is substrate, 3 is electrode, 4 is roller conveyer-electrode and 5 is plasma).

reactors are used in industrial production. It allows to increase production rate and to avoid reciprocal contamination of layers. Multi-chamber reactor with semi-continuous process of a–Si:H solar cells production is shown schematically in Fig. 5.9 (Vikhrov and Bodygin, 1994).

There is a number of consecutive chambers with independent pumping-out systems. Chambers are separated one from another and from external atmosphere by vacuum seals (1). Substrates (2) transfer from one chamber to the other by roller conveyer (4) which simultaneously serves as the lower electrode of radio-frequency system. First, the substrate is placed into inlet chamber. Inlet vacuum seal shuts and air evacuated from chamber down to ultimate vacuum. Further the next vacuum seal opens and substrate transfers to the second chamber which is already evacuated down to ultimate vacuum. Vacuum seal shuts after substrate transference and *p*-type a–Si:H layer deposition process starts after gas mixture fills up the chamber and glow discharge initiates. At the same time another substrate is placed to inlet chamber. When deposition of *p*-type layer process is finished substrate transfers to the third chamber to deposit *i*-type layer. Further the process continues in the same sequence up to *p-i-n* structure preparation. It is necessary to note that industrial reactors have larger number of vacuum chambers for preparation of contact layers, protection layers and etc. Thus reactors allow to prepare finished solar cells.

Schematic illustration of a multi-chamber reactor with continuous (roll to roll) process of *p-i-n* a–Si:H solar cells production is shown in Fig. 5.10 (Popov, 2008). The reactor includes a number of consecutive chambers with independent pumping-out systems. Stainless steel foil (3) is used as substrate. It rewinds continuously from feeding roll (4) in the first chamber to receiving roll (5) in the last

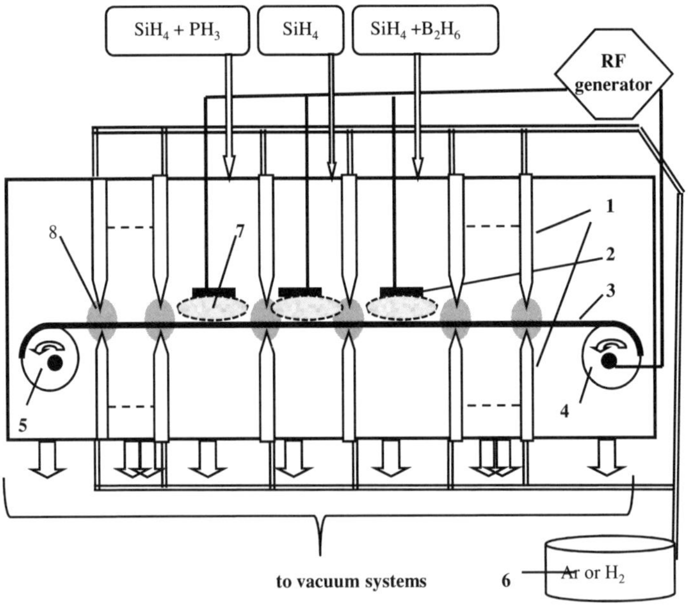

Figure 5.10. Schematic illustration of multi-chamber reactor with continuous (roll to roll) process of *p-i-n a*–Si:H solar cells production (1 is gas seal, 2 is electrode, 3 is stainless steel substrate, 4 and 5 are feeding and receiving rolls, 6 is gas feeding system, 7 is plasma and 8 is area of higher pressure).

one and passes through all chambers. Chambers are separated one from another by gas seals (1). Gas feeding system (6) creates areas of higher pressure (8) in gas seals. It prevents chambers from mutual contamination. The deposition of contact layers, a–Si:H layers with different types of conductivity, upper electrode and protection layer is carried out in the chambers and finished solar cells are obtained on receiving roll.

As it was mentioned above glow discharge decomposition method is characterized by relatively low film growth rate (about 1 μm/hour). If the film thickness less 1 μm is necessary for device application (thin film transistors or solar cells for example) such growth rate does not hinder the technological process. But film thickness up to 30 μm is required for some applications (electro-photographic drums for example). In this case it is necessary to increase growth rate by at least a factor of 10. It is certainly possible to increase radio-frequency power. Such solution ensures acceleration of film growth but at the same time it leads to an increase of localized state concentration in the gap and to sharp deterioration of film properties (photoconductivity decreasing in particular).

Substitution of silane for polysilanes (Si_2H_6 or Si_3H_8) increases growth rate by up to 20 times within the same technological conditions. Film properties remain practically the same as well (Spear and Le Comber, 1984). Dependencies of conductivity activation energy on substrate temperature in the case of a–Si:H films prepared from SiH_4 and Si_2H_6 are shown in Fig. 5.11. Radio-frequency power

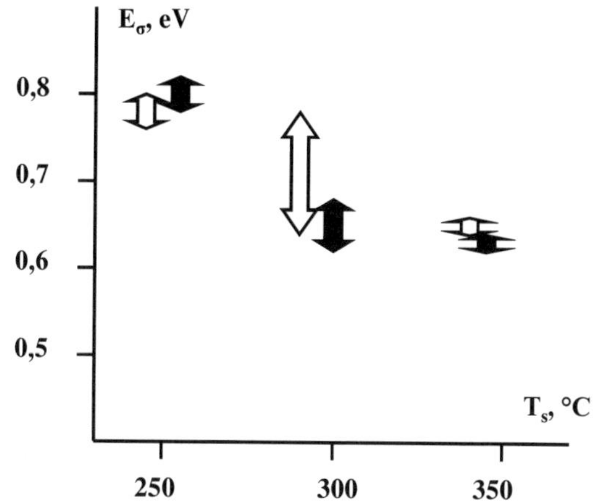

Figure 5.11. Dependences of conductivity activation energy on substrate temperature in the case of a Si:H films prepared from SiH_4 (open marks, thickness is 1 μm) and Si_2H_6(solid marks, thickness is 20 μm).

and preparation time were the same in both cases but film thickness is 1 μm in the former case and 20 μm in the latter case. One can see from the figure that differences lay within the limits of experimental errors. The reason for a–Si:H film growth rate acceleration is in the following. The limiting reaction stage that determines film growth rate is generation of SiH_2 radicals due to decomposition of silane molecules In the case of SiH_4 decomposition reaction is given by

$$SiH_4 \longrightarrow SiH_2 + H_2 \tag{5.2}$$

and in the case of disilane decomposition reaction is given by

$$Si_2H_6 \longrightarrow SiH_2 + SiH_4 \tag{5.3}$$

Activation energy of the latter is considerably less. For example, rate of reaction (5.3) exceeds reaction (5.2) rate by 100 times at the temperature of 400 °C. Therefore higher film growth rate in the case of polysilane decomposition is explained by the increase of SiH_2 radical generation.

Thus substitution of silane for polysilanes ensures acceleration of a–Si:H film growth rate up to 20 μm/h. However polysilanes costs are significantly higher than SiH_4. Moreover polysilanes are unstable compounds. That is why, as a rule, the synthesis of polysilane is carried out just before the film growth process, usually in the same reactor. This leads to more complexity in the production equipment and to the rise of the production costs.

The other problem of the devices based on a–Si:H development is connected to impossibility of low resistance material preparation. Heavy doped low resistance n^+ and p^+ layers are usually used to create ohmic (non-rectifying) contact in semiconductor devices. However, as was mentioned in Sec. 4.1, it is impossible to prepare hydrogenated amorphous silicon with electrical conductivity more than 10^{-2} Ohm^{-1} cm^{-1} and conductivity activation energy less than 0.2 eV because of the high density of localized states in the band tails. Thin layers of hydrogenated microcrystalline silicon (μc–Si:H) can be used to create ohmic contacts in devices based on a–Si:H because conductivity of μc–Si:H is higher than a–Si:H conductivity by three orders of magnitude. Hydrogenated microcrystalline silicon is prepared by the same method of glow discharge silane decomposition and in the same reactors by means of process characteristic changes.

In the general case, reaction of silane decomposition in glow discharge is expressed by

$$SiH_4 \leftrightarrow SiH_x + (4-x)H \leftrightarrow (Si + xH)_{sol} + (4-x)H_{pl} \tag{5.4}$$

where index 'sol' means solid phase and index 'pl' means plasma. Direct reaction (from left to right) leads to silane decomposition and a–Si:H film growth. Reversible reaction (from right to left) results in film erosion (etching of film in plasma). In the case of hydrogenated amorphous silicon preparation direct reaction predominates because the system is far from chemical equilibrium. Proximity to chemical equilibrium leads to slower film growth rate and to increasing mobility of condensed atoms on the substrate. In turn this creates conditions for crystalline nucleus formation.

Conditions close to chemical equilibrium are ensured by silane dilution with hydrogen up to 2–3% silane and increasing glow discharge power by 5–10 times. As a result, the film growth rate falls by a factor of about 10. It is evidence of reaction shift towards chemical equilibrium. Diffraction patterns of silicon films prepared under these conditions show amorphous-crystalline state of material (Spear and Le Comber, 1984). Content of crystalline phase in the film is 65–97% depending on preparation conditions (Fig. 5.12). Crystallite sizes range from 50 Å to 80 Å.

Dependence of hydrogenated microcrystalline silicon conductivity on doping by phosphorus and boron is shown in Fig. 5.13 (Spear and Le Comber, 1976). One can see from the figure that conductivity of undoped μc–Si:H is approximately 10^{-2} Ohm^{-1}cm^{-1}. It is higher than undoped a–Si:H conductivity by six orders of magnitude. The values of μc-Si:H film conductivity 20 Ohm^{-1}cm^{-1} and 8 Ohm^{-1}cm^{-1} have been obtained by PH_3 and B_2H_6 doping correspondingly. This allows to create ohmic contacts in devices based on hydrogenated amorphous silicon.

Investigations of hydrogenated amorphous silicon stimulated search for new materials with better properties. One of these attempts consisted in substitution of hydrogen with fluorine because energy of Si–F bond is significantly more than energy of Si–H bond. It had hopes to decrease localized state density in the gap of material and to minimize gas effusion at elevated temperatures. A series of investigations has been carried out with substitution of SiH_4 with SiF_4 (Madan, 1984). Various gas combinations were used: SiF_4, SiF_4 + Ar, SiF_4+ H_2, SiF_4 + SiH_4.

However total replacement of hydrogen with fluorine gives negative result. Localized state density in a–Si:F films increases in comparison with a–Si:H films. The reason may be in the big value of fluorine electronegativity (3.9). (Silicon

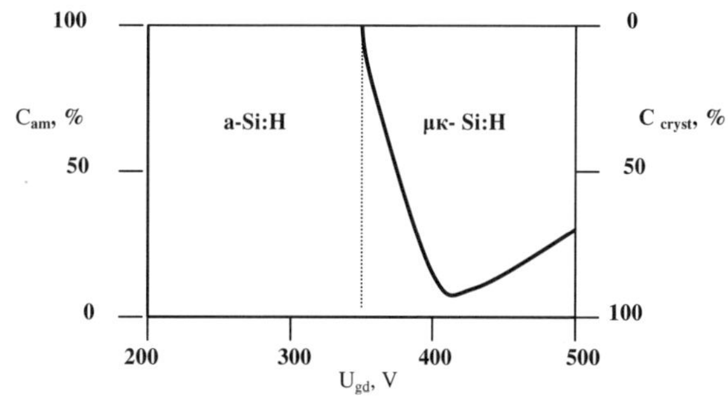

Figure 5.12. Dependence of amorphous (C_{am}) and crystalline (C_{cryst}) phases content in silicon film on glow discharge voltage (U_{gd}).

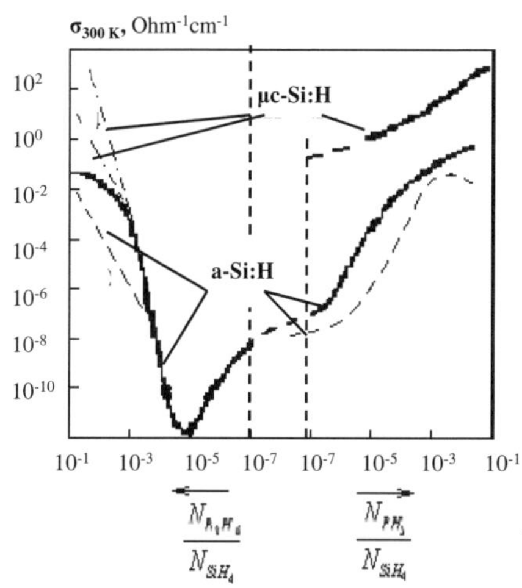

Figure 5.13. Electrical conductivity of a–Si:H and μc–Si:H dependences on doping by phosphorus and boron level.

electronegativity equals 1.8 and hydrogen electronegativity equals 2.1). It leads to significant ionicity of Si–F chemical bonds, polarization of neighboring Si–Si chemical bonds and to growth of localized state density in the gap as a result. At the same time combined application of fluorine and hydrogen containing gases brings a positive result. Silicon films prepared from SiF_4 and H_2 mixture (12–20% hydrogen) have lower localized state density in the upper part of forbidden gap in comparison with a–Si:H films. Dependencies of electrical conductivity and activation energy of conductivity on doping level of a–Si:F:H films are shown in Fig. 5.14 (Madan and Ovshinsky, 1980). The curves of a–Si:H films are shown in the figure as well. Characteristic feature of fluorinated films is higher conductivity value in the case of heavy doped *n*-type samples. The following explanation may be provided for this fact. Films with high fluorine content are inclined to crystallization because of chemical bond ionicity increasing. That leads to silicon

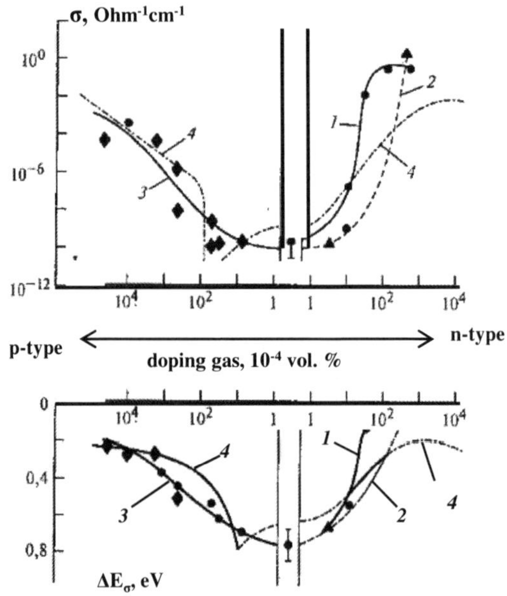

Figure 5.14. Dependences of electrical conductivity (σ) and activation energy of conductivity (ΔE_{σ}) on doping level of a–Si:F:H films doped by arsenic (1), phosphorus (2) and boron (3) and a–Si:H films(4).

microcrystalline phase formation even at the normal regimes of film preparation and allows to create ohmic contacts to n-type a–Si:H.

5.3 PREPARATION OF A^{IV} B^{IV} ALLOYS ON THE BASE OF HYDROGENATED AMORPHOUS SILICON

Combination of materials with different values of forbidden gap allows to create new highly effective optoelectronic and photoelectric devices. Compounds $A^{III}B^{V}$ and $A^{II}B^{VI}$are used for this purpose in the case of crystalline semiconductors. Forbidden gap value changes in chalcogenide glassy semiconductors can be obtained by variation of material chemical composition. In the case of tetrahedral bonded amorphous semiconductors alloys silicon and other elements of IV group of periodic table are used to produce variation of the material's gap value.

Silicon forms solid solutions with other elements of IV group (germanium, carbon, nitrogen). These elements or their compounds with silicon have the value of gap smaller (germanium) or bigger (SiC and SiN_4) than value of silicon gap. Therefore admixture of gaseous compounds of germanium, carbon or nitrogen with hydrogen to silane allows to vary the value of forbidden gap of material within a wide range (Table 5.1).

A–$Si_{1-x}Ge_x$:H films with x value from 0 to 1 are prepared by glow discharge decomposition of gas mixture SiH_4 + GeH_4 + H_2. Infra-red spectroscopy study

Table 5.1 Forbidden gap value of A^{IV}–B^{IV} materials.

Material	ΔE, eV
a–Si:H	1.8
a–$Si_{1-x}Ge_x$:H	1.1–1.8
a–$Si_{1-x}C_x$:H	1.8–2.6
a–$Si_{1-x}N_x$:H	1.8–4.0

shows presence in the films of both chemical bonds Si–H and Ge–H. However, the relative concentration of Si–H bonds exceeds relative concentration of Ge–H bonds by approximately the factor of 3:[Si–H]/[Si] = 0.11 and [Ge–H]/[Ge] = 0.04. This indicates that hydrogen atoms are more effective in passivating silicon dangling bonds in comparison with germanium atoms. At the same time relative concentration of germanium dangling bonds (non-paired spin concentration [Ge↑]/[Ge]) decreases more quickly than concentration of Ge–H bonds ([Ge–H]/[Ge]) grows. It means that hydrogen atoms not only passivate dangling bonds but also decrease their concentration due to structural relaxation. Dependence of forbidden gap value on component ratio in the films of a–Si_{1-x} Ge_x:H system is shown in Fig. 5.15. One can see that silicon-germanium ratio changes lead to linear variation of material gap value in the limits from 1.1 to 1.8 eV.

Films of a–Si_{1-x} C_x:H system are prepared by glow discharge decomposition of mixture silane and methane (CH_4) or silane and ethylene (C_2H_4). Dependence of material forbidden gap value on carbon content (fig. 5.16) has more complicated form in comparison with materials of a–$Si_{1-x}Ge_x$:H system and changes under variation of film preparation conditions. The reason lies in the possibility of carbon atoms to have various electron state hybridizations and to form different structural units.

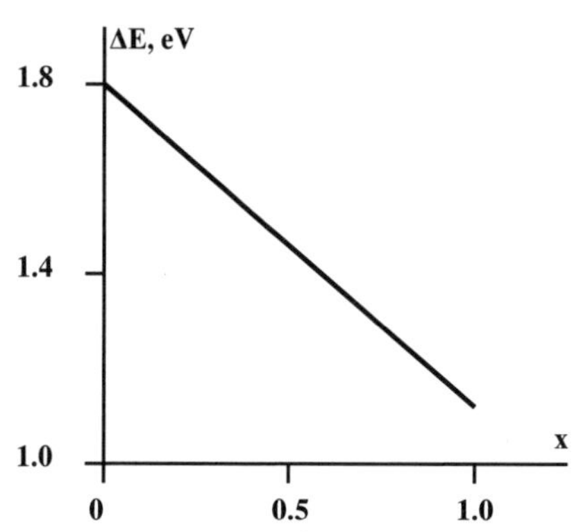

Figure 5.15. Dependence of forbidden gap value on component relation in the films of a–Si_{1-x} Ge_x:H system.

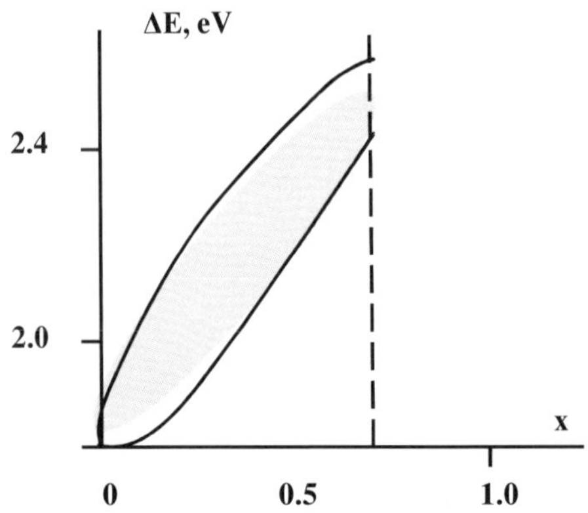

Figure 5.16. Dependence of forbidden gap value on component relation in the films of a–Si_{1-x} C_x:H system.

It is necessary to note that the increase in the number of components leads to the rise in defect concentration in the film. Therefore, in comparison with a–Si:H, the density of localized states in the a–Si_{1-x} C_x:H and a–Si_{1-x} Ge_x:H systems is higher. Thus the change the material's gap value leads to degradation in the material's optoelectronic quality.

5.4 PREPARATION OF HYDROGENATED AMORPHOUS SILICON FILMS BY CHEMICAL VAPOR DEPOSITION (CVD) METHODS

Thermal decomposition (pyrolysis) of silicon-contained gases (SiH_4, $SiCl_4$) is used widely in single crystal silicon device production for epitaxy layer preparation. It would be convenient to apply this well known technology to hydrogenated amorphous silicon film preparation. Thermal CVD reactor consists of a quarts tube located in a furnace. Silicon-contained gas (silane, for example) is thermally decomposed in tube and silicon atoms deposit at the substrate. Film growth rate depends exponentially on temperature. In the case of epitaxy process temperature range 900–1000°C is usually used. However, these values are not applicable to amorphous silicon film growth because crystallization temperature of amorphous silicon lies in the limits of 670–700°C. Therefore temperature between 550 and 650°C is usually used in the latter case. It ensures amorphous silicon film growth rate of about 100Å/min. This value is acceptable for some applications. But there is one more problem. Relatively high temperature of deposition leads to sharply decreasing concentration of hydrogen in a–Si:H film and to growing localized state density as a result.

Various ways were used to overcome this problem (Kaplan, 1984). Additional film treatment in hydrogen plasma (post-hydrogenation) is one of for examples. It increases hydrogen concentration in the films but material quality does not reach the level of films prepared by glow discharge decomposition method. The other way is photo-CVD method which employs combined thermal and ultraviolet

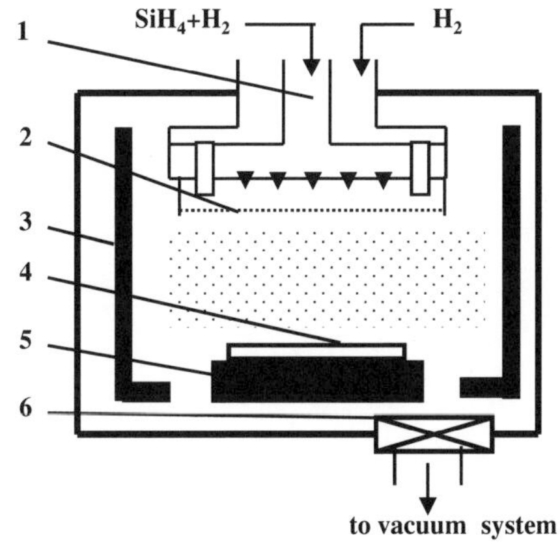

Figure 5.17. Schematic illustration of hot wire chemical vapor deposition reactor (1 is gas feeding system, 2 are hot wires, 3 are hot walls, 4 is substrate, 5 is pedestal and 6 is gas feeding system).

radiation treatment to decompose silane. This allows to decrease decomposition process temperature and to raise hydrogen concentration in the films. The same result can be reached by substitution of silane with disilane (see Sec. 5.2). However, significantly more complex production equipment is needed in both cases.

Another modification of thermal CVD method is called catalytic chemical vapor deposition or hot wire chemical vapor deposition (Matsumura *et al.*, 2003). Gas mixture passes through tungsten or tantalum filaments heated to 1600–1800°C (Fig. 5.17, Sunayama *et al*, 2003) and silane decomposition takes place without heating the whole chamber space. Substrate temperature is kept in the limits of 250–350 °C that ensures sufficient hydrogen concentration in growing film and low density of localized states in the material. High quality hydrogenated amorphous silicon films have been prepared by this technique. Characteristics of thin film transistors and solar cells on the base of these films are not inferior to the devices on the base of films prepared by glow discharge decomposition method.

5.5 PREPARATION OF HYDROGENATED AMORPHOUS SILICON FILMS BY RADIO-FREQUENCY SPUTTERING METHOD

Hydrogenated amorphous silicon films are prepared by silicon target sputtering in argon-hydrogen atmosphere. Hydrogen atoms interact with sputtered silicon particles and are incorporated in growing film (reactive sputtering method). Hydrogen concentration depends on hydrogen partial pressure, on discharge power and substrate temperature. Forbidden gap value increases with increasing hydrogen concentration (Fig. 5.18).

Localized state density in a–Si:H films prepared by sputtering method is higher in comparison with films prepared by silane decomposition method. However, the former provides possibility of film preparation with resistivity of up to

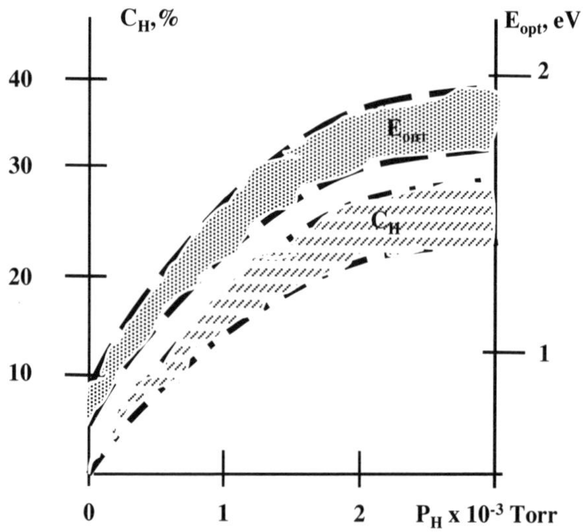

Figure 5.18. Hydrogen concentration (C_H) and optical gap value (E_{opt}) of a–Si:H dependences on hydrogen partial pressure (ranges of values correspond to different substrate temperature and discharge power).

10^{13} Ohm cm. That is greater by the factor of 10^2 than maximal resistivity values of films prepared by silane decomposition method. High resistivity is a necessary condition for some application (electro-photographic drums, for example, see Chapter 6).

There are several ways of material doping in the case of sputtering method (Thompson, 1984). Firstly phosphine or diborane may be added to

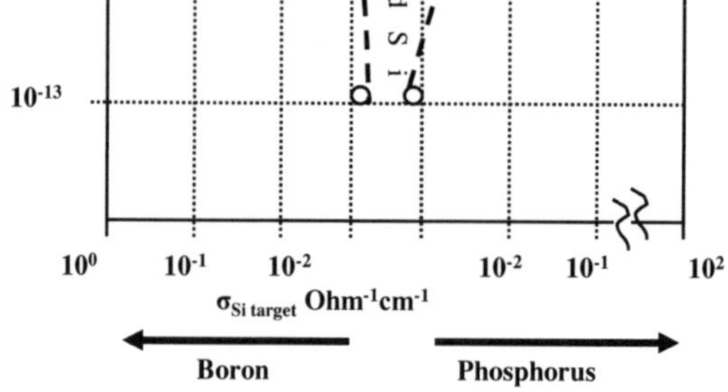

Figure 5.19. Dependence of a–Si:H film conductivity on silicon target conductivity.

argon-hydrogen atmosphere in the chamber. Secondly sputtering of doped silicon target leads to doping growing film. Film conductivity changes by 5–7 orders of magnitude depending on target conductivity (Fig. 5.19). Thirdly thermal evaporation of phosphorus or boron containing compounds is carried out during film preparation process.

Summing up one may see that there are quite a few methods of disordered semiconductor film preparation. These methods have various merits from point of view of prepared film and device quality as well as complexity and cost of manufacturing. The specialist proficiency consists in the choice of method which ensures appropriate device quality at reasonable manufacture cost.

Chapter Six

Optical Information Storage and Transmission Devices

Disordered semiconductors are widely used as optical information storage and transmission devices as well as high resolution photo- and electronic resists (Fig. 6.1).

There are four different groups of optical information recording methods. All methods are based on creating the image on the surface of semiconductor layer. The difference in the methods is in how this image is recorded. The first group includes methods based on the representation by electrical charge distribution on the surface of semiconductor layer. The second group includes methods based on representation of the image by inducing structural changes in non-crystalline phase under the light exposure, leading to local changes of the material's optical and chemical properties. The third group includes methods based on representation of the image by inducing interaction between chalcogenide glassy semiconductor and metal under the light exposure, leading to local changes of the material's optical and chemical properties, similar to group two described above. Last but not least, the fourth group includes methods of representation using light-induced first-kind phase transitions. These include local crystallization of non-crystalline phase or local amorphization of crystalline phase or local material evaporation under light irradiation. While all methods and optical recording devices based on disordered semiconductors are characterized by reusability, real-time recording and high resolution, they are also marked with relatively low photosensitivity (Table 6.1).

6.1 DEVICES BASED ON CHARGE PATTERN RECORDING

The first group (Fig. 6.1) includes devices with the recording medium and photosensitive non-crystalline semiconductor layer (photoreceptor) on conductive substrate. Layer surface is charged to some initial potential via corona discharge

Disordered Semiconductors: Physics and Applications by Anatoly Popov

www.panstanford.com
978-981-4241-76-2

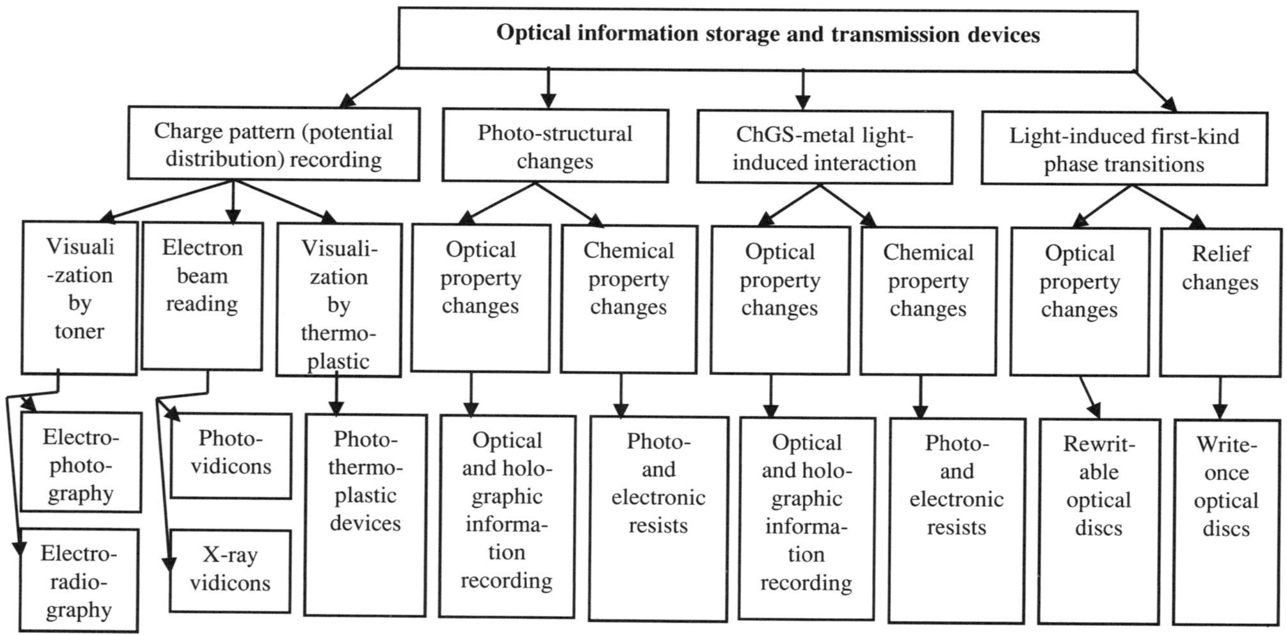

Figure 6.1. Optical information storage and transmission on the base of disordered semiconductors(ChGS is chalcogenide glassy semiconductors).

Table 6.1 Sensitivity and resolution of disordered semiconductor devices and human eye.

Methods	1st group	2nd group	3rd group	4th group	Human eye
Sensitivity, cm^2 /J	$10^5 - 10^7$	$10^{-2} - 10^1$	$10^2 - 10^3$	$10^{-1} - 10^2$	10^{10}
Resolution, mm^{-1}	$10 - 500$	10^3	$(5 - 10) \cdot 10^3$	$\geq 10^3$	10^1

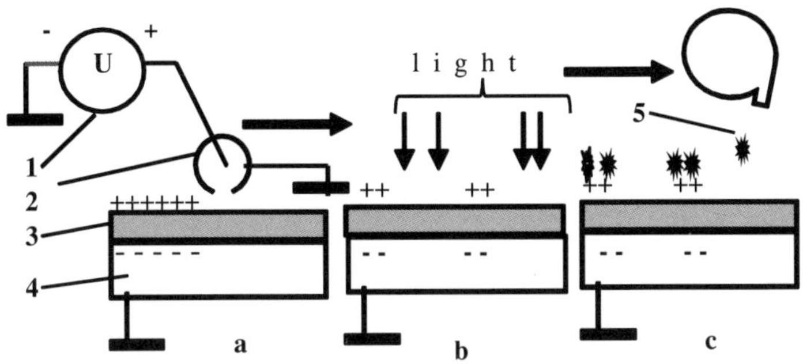

Figure 6.2. Creation and visualization of latent electrostatic image in electro-photography (see text):1 is voltage source, 2 is korotron, 3 is semiconductor layer, 4 is conductive substrate and 5 are charge toner particles.

(in the case of electrographic or phototermoplastic devices, Fig. 6.2(a)) or via electron beam (in the case of vidicons, Fig. 6.12). Then an optical image is projected onto the charge surface. Surface potential is reduced in the regions exposed to light. Potential distribution that replicates the image pattern (latent electrostatic image) is formed due to semiconductor layer photoconductivity (Fig. 6.2(b)). Visualization of the latent electrostatic image or reading of potential distribution is realized by charged toner particles in the case of electrography (Fig. 6.2(c)), by heated thermoplastic film in photothermoplastic devices (Fig. 6.11) or by electron beam scanning in vidicons (Fig. 6.12).

The main function of semiconductor layer is to convert optical image to electrostatic image. Therefore semiconductor layer has to have high value of photoconductivity in visible spectrum to create sufficient differences in surface potential between light and dark areas during exposure time. At the same time semiconductor layer has to have high resistance to maintain differences in surface potential during processing time. Time of exposure and reading of potential distribution is equal to TV frame time (0.04 sec) in vidicons. Semiconductor layer resistivity has to be no less than 10^{11} Ohm cm to maintain potential distribution during this time. Even higher resistivity of semiconductor layers is necessary in the case of electro-photography and photo-thermoplastic devices.

The requirements mentioned above contradict one another. Semiconductor conductivity (and resistivity) is determined by concentration and mobility of charge carriers

$$\sigma = 1/\rho = qn\mu_n + qp\mu_p \tag{6.1}$$

where q is electron charge, n and p are electron and hole concentrations and μ_n and μ_p are electron and hole mobility. The minimum concentration of charge carriers in intrinsic semiconductor is determined by forbidden gap value

$$n_i = p_i = A \quad \exp[-\Delta E/(kT)] \tag{6.2}$$

where ΔE is forbidden gap value, k is Boltzmann's constant and T is temperature. Thus it is desirable to use wide-band semiconductors to increase layer resistance.

On the other hand maximum photoconductivity of material takes place when photon energy is equal to semiconductor gap value (see Sec. 3.4). One usually records the image using visible light. Since visible spectrum of electromagnetic radiation lies in the wave length diapason from 0.40–0.75 μm semiconductor gap value has to be equal 3.1–1.7 eV correspondingly ($\lambda = 1.234/\Delta E$) to ensure high photoconductivity. However traditional crystalline semiconductors have insufficient resistivity at the given gap values to maintain potential distribution. Use of disordered semiconductors instead of monocrystalline ones solves this contradiction because the former have resistivity 4–5 orders of magnitude higher at the same gap values. The reason for this lies in considerably lower mobility of charge carriers in the case of disordered semiconductors (see Sec. 3.2).

6.1.1 Electro-photographic Process

The first electro-photographic copy was created in 1938 by C. Carlson (Carlson, 1942). Twenty one years later, in 1959, the first electro-photographic copying machines (model 914) were manufactured and gained wide popularity. Since then electro-photographic process became the most widespread method of copying. Today, there are about four billion copies made daily in the world. Besides that, the electro-photographic process is used actively in other fields: printers, fax machines and etc. More than five millions copiers and more than fifteen millions printers are produced annually.

Schematic illustration of the most widespread electro-photographic rotary copier is shown in Fig. 6.3 (Popov, 1997).

Photosensitive electro-photographic drum consists of high resistivity disordered semiconductor layer on cylindrical conductive substrate. The drum rotates around its own axis at a constant speed. All process steps are performed on the circumference of the drum. Electrical charge is sprayed on the surface of semiconductor layer via corona discharge 1. Double charge layer is formed by surface charges and screening charges in conductive substrate as a result (stage *a*). Part of image pattern in the form of narrow strip is projected onto charged drum surface through the optical system 2. Due to coordinated movement of copied pattern and drum the projection of the former transfers completely on the surface of semiconductor layer. At the illuminated regions surface potential is reduced by virtue of semiconductor photoconductivity. Discharge current flows normally to the surface and surface potential distribution or latent electrostatic image replicates the image pattern (stage *b*). Image development is realized in chamber 3.

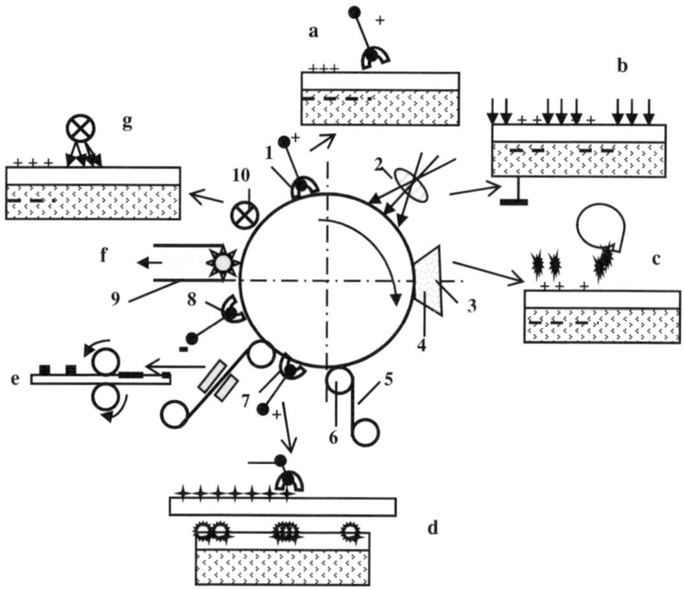

Figure 6.3. Schematic illustration of electro-photographic rotary copier(see text).

Mixture of toner particles and cattier beads 4 spreads over drum surface. Toner particles are charged by triboelectrical effect and are attracted to dark surface areas by Coulomb interaction. Visible image corresponding to the latent electrostatic image is formed as a result (stage *c*). Developed image transfers on paper 5 (or another carrier) through pressing it to semiconductor layer surface by pinch rollers 6 and charging it via corona discharge 7. Charged toner particles displaces on the paper under electric field action (stage *d*). Print fixing is carried out by heat, heat and pressure or laser treatment (stage *e*). The electro-photographic copy is completed. However, it is necessary to prepare the surface of semiconductor layer for the next cycle. That is to clean the surface of residual toner particles and electrical charges. To facilitate cleaning electrical charge of opposite sign is sprayed on the surface via corona discharge 8. Residual toner particles are mechanically removed 9 (stage *f*). Finally light irradiation 10 is carried out to delete residual electrical charges by photoconductivity (stage *g*). The drum is ready for the next cycle.

Printers differ from copiers only by the method of latent electrostatic image creation (stage *b*). The latent image is formed line by line by selective exposure laser beam (laser printer) or line of light-emitting diodes (LEDs printer). All the other stages remain unchanged.

The main performance targets of electro-photographic machines are copying rate, optical density of the copy, drum durability, original and copy sizes, the possibility of scale variation, weight-size and power consumption characteristics. Most of them are determined by properties of semiconductor material which is used in electro-photographic drum.

The two major and contradictory requirements for these materials are high photoconductivity in visible spectrum and high resistivity in the dark. This contradiction has been resolved through the use of disordered semiconductors (see above). The other requirements for semiconductor material for electro-photography include: crystallization resistance, mechanical strength and stability of properties under influence of high electric fields, ionic currents and toner particles. The current trend to reduce the size of electro-photographic machine results in deterioration of heat removal. As a result the requirements for temperature stability of semiconductor materials are increasing.

Based on the above mentioned requirements five groups of semiconductor materials found application as photoreceptors in electro-photography:

- polycrystalline semiconductor (zinc oxide) in the dielectric medium (this was the first material to be used, but it is no longer used),
- glassy selenium and selenium alloys ,
- two- and multi-component chalcogenide glassy semiconductors,
- organic semiconductors,
- hydrogenated amorphous silicon and its alloys.

Electro-photographic photoreceptors are produced on rigid (cylindrical or planar) or flexible substrates. In all cases it includes several layers that perform different functions, namely (Fig. 6.4):

- the layer of carrier generation (2) where absorbed photons produce electron–hole pairs,
- transportation layer (3) through which photo-generated carriers move toward the substrate under electrical field action,

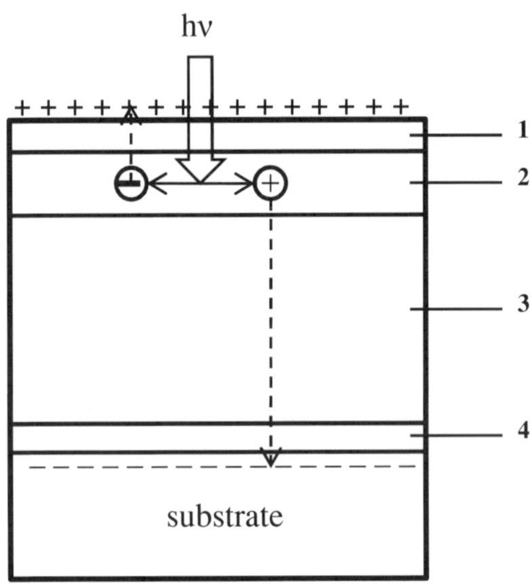

Figure 6.4. Schematic illustration of photoreceptor(see text).

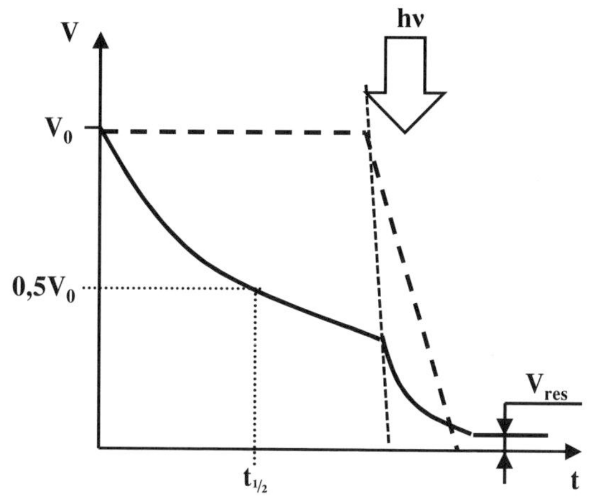

Figure 6.5. Dark and light discharge characteristics of ideal and real photoreceptors(V_0 is initial surface and V_{res} is residual voltage.

- the layer creating a blocking contact between semiconductor and conductive substrate (4) which prevent carrier injection from the substrate,
- protective layer (1) on the photoreceptor surface.

It should be noted that presence of all these layers is not necessary. In some cases several functions are combined in a single layer.

Dark and light discharge characteristics of ideal and real photoreceptors are shown in Fig. 6.5. There is no initial surface voltage decay in the dark in the case of an ideal photoreceptor. With light turned on, the discharge rate is directly proportional to the light intensity and surface voltage decreases down to zero.

In practice, the situation differs significantly. Dark voltage decay is present because of the finite value of photoreceptor's resistance. Dark discharge rate is characterized by the time of initial surface voltage reduction by half ($t_{1/2}$) or by the slope of discharge curve at $t = 0$. Light discharge rate depends on surface potential, photon energy and is not directly proportional to light intensity. Voltage across the photoreceptor does not drop to zero after exposure due to trapping of charge carriers on the deep localized states in the gap. Residual potential (R_{res}) remains even after photoreceptor's complete light discharge.

Electro-photographic sensitivity (S) can be characterized as light discharge rate (dV/dt) or as exposure needed to reduce initial voltage to a predetermined value. The latter is more convenient in practical application. However, it is necessary to specify the relative value of initial voltage reduction in this case: $[V_0 - \Delta V]/V_0$, where V_0 is initial voltage and ΔV is reduction of voltage as a result of exposure (contrast potential). Therefore sensitivity subscript 0.5 ($S_{0.5}$) means that measurements were carried out at the exposure which reduces surface voltage by half and $S_{0.1}$ means reduction of surface voltage by 90%. Energy units of sensitivity (square meters per joule) are used if spectral characteristics of photoreceptors are analyzed. In this case sensitivity is inverse value of radiation energy incident upon surface unit. However if the spectral composition of light is constant the sensitivity is determined as inverse value of light exposure required for a given surface voltage reduction ($lux^{-1}sec^{-1}$).

6.1.2 Semiconductor Materials for Photoreceptors

Selenium photoreceptors. For a long time, selenium photoreceptors were the most widespread, mainly due to the match of their main characteristics to the requirements of electro-photographic process. Selenium drum manufacturing process and technology are described in Chapter 5. Selenium photoreceptors are usually produced on duralumin substrate. The substrate is anodized to create a blocking contact. The photoreceptor itself is a multilayer system that includes the following parts: blocking oxide film that is formed on the surface of conductive substrate; crystalline (trigonal) selenium layer that adjoins the oxide film; non-crystalline selenium transportation layer. The next layer is a carrier generation layer, which is characterized by somewhat ordered non-crystalline selenium atomic structure created with the help of special deposition conditions. A protective layer on the photoreceptor's surface is formed by treatment in gas mixtures or by thin dielectric film deposition (organic films or diamond-like amorphous carbon films).

Spectral dependence of selenium photoreceptor sensitivity is shown in Fig. 6.6. One can see that there are two maxima of photosensitivity. The first maximum at the wave range 430–470 nm is defined by carrier generation in non-crystalline selenium. The second maximum at the wave range 720–750 nm is defined by carrier generation in crystalline selenium layer adjoined to substrate. This peak appears because of weak absorption of long-wave photons in non-crystalline selenium. However the intensity of this peak is by an order of magnitude lower comparing to the intensity of the first peak.

It is possible to control some characteristics of selenium photoreceptors (crystallization resistance and mechanical strength, for example) by means of structural

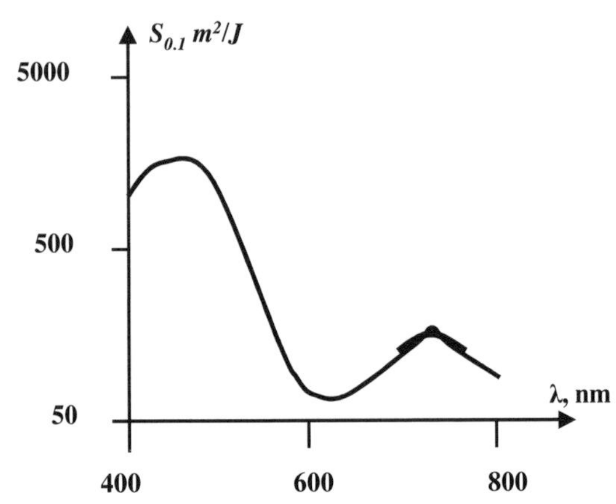

Figure 6.6. Spectral dependence of selenium photoreceptor senstivity.

modification at the level of medium range order (see Chapter 4). On the other hand, this method does not ensure broadening or shift of photosensitivity spectral range because the latter depends on forbidden gap value. Modifications to the photosensitivity spectra are done by admixture addition to selenium.

Admixtures of arsenic, antimony, germanium, bismuth and tellurium lead to a shift of selenium photoreceptor sensitivity to long-wave spectrum range. At the same time, these admixtures provoke deterioration of other characteristics: decrease of carrier drift mobility and of the absolute value of photosensitivity, increase of dark discharge rate and of residual potential value, change of crystallization activation energy. Bismuth and tellurium admixtures offer the best application-tailored compromise (Popov, 1997).

Photoreceptors of $Se_{100-x}Bi_x$ system are usually prepared by thermal co-evaporation (see Chapter 5) of selenium and bismuth. X-ray and electron diffraction analysis does not display crystalline phase in $Se_{100-x}Bi_x$ films up to $x = 30$ (Bowman and Schottmiller, 1968, Schottmiller, Bowman and Wood, 1968). Bismuth admixture leads to broadening of the valence band and to a reduction of selenium forbidden gap value due to formation of selenium–bismuth bonds which have lower energy in comparison to selenium–selenium bonds (Popov, Mikhalev, Karalynets and Vasillyeva, 1991). Reduction of gap value leads to decreasing resistivity of material. $Se_{100-x}Bi_x$ film resistivity as a function of bismuth concentration is shown in Fig. 6.7(a). Thirty percent of bismuth decreases material resistivity by 11 orders of magnitude (Schottmiller, Bowman and Wood, 1968). The most considerable resistivity changes take place at low bismuth concentration.

Spectral characteristics of photoreceptor sensitivity change as well. $Se_{100-x}Bi_x$ photosensitivity maximum as a functiom of bismuth concentration is shown in Fig. 6.7(b). One can see that the photosensitivity maximum shifts to the long wave range with bismuth concentration increasing up to 5 atomic precent. Subsequent

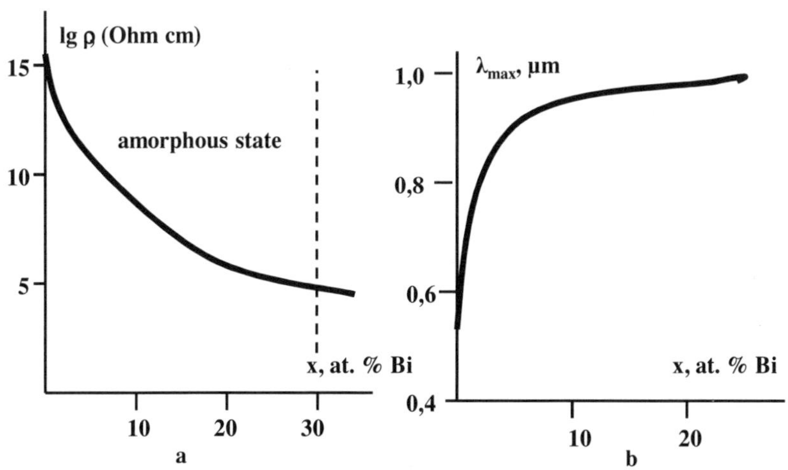

Figure 6.7. Concertration dependences of $Se_{100-x}Bi_x$ film resitivity (a) and wave length corresponded to photosensitivity maximum (b).

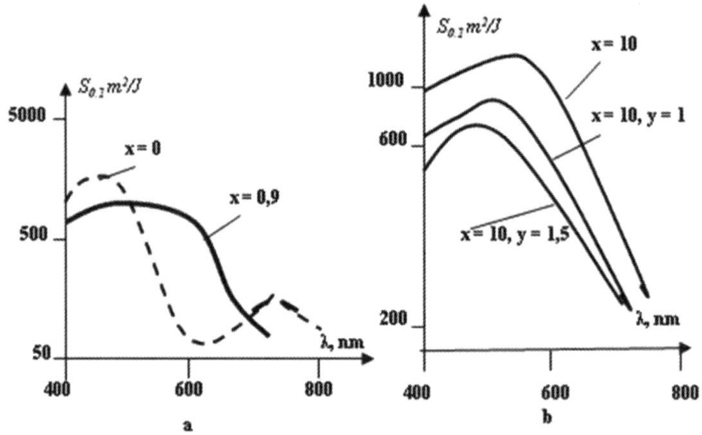

Figure 6.8. Spectral dependences of $Se_{100-x}Bi_x$ (a) and $Se_{100-x}Te_x$, $Se_{100-x-y}Te_xP_y$ (b) photorecptors senstivity.

increase of bismuth concentration has weak influence on photosensitivity maximum position.

Thus, from the point of view of photosensitivity maximum shift it is not really useful to increase bismuth concentration to more than 5 atomic percent. The material resistivity falls down to 10^{10} Ohm cm at bismuth concentration of 5 atomic percent. This is too low a value to maintain differences in surface potential during processing time. Hence bismuth concentration from 0.5 to 2 atomic percent is used in photoreceptors of $Se_{100-x}Bi_x$ system. Such systems have spectral dependence of photosensitivity shifted to the long-wave part of spectrum up to the red light region (Fig. 6.8(a)).

It should be noted that bismuth admixture sharply decreases drift mobility of holes (from 0.16 cm^2 /V sec in pure non-crystalline selenium to 10^{-3} cm^2 /V sec) and increases residual potential value.

In general, bismuth admixture to selenium leads to:

- shift of photoreceptor sensitivity to the long-wave spectrum region,
- increase of dark discharge rate
- increase of residual potential value.

In contrast to bismuth, admixture of tellurium to selenium has no influence on carrier drift mobility. This leads to broadening of photosensitivity spectral dependence towards long waves up to 800 nm at 10 atomic percent of tellurium. At the same time it leads to inadmissible rise of photoreceptor dark discharge rate due to resistivity decrease.

There are two ways to address this problem (Popov, 1997). The first is limiting the tellurium concentration at the level of 1–3 atomic percent. The second involves simultaneous admixture of tellurium and other elements that reduce dark discharge rate. For example, phosphorus admixture to Se_{100-x} Te_x layers leads to a decrease of dark discharge rate by order of magnitude. However, at the same time it degreases photosensitivity by 30–40% as well (Fig. 6.8(b)).

The last but not the least of the shortcomings of Se_{100-x} Te_x photoreceptors are lower crystallization activation energy and, as a result, lower durability.

Arsenicum tri selenide photoreceptors. Glassy arsenicum tri selenide has layered polymeric structural network, consisting of pyramidal blocks $AsSe_{3/2}$ (see Chapter 2). This type of structure facilitates the through passage of current carriers and provides high discharge efficiency with light exposure. Photosensitivity of As_2 Se_3 photoreceptors under exposure of incandescent lamp light considerably exceeds selenium photoreceptor photosensitivity, mostly due to broadening of spectral sensitivity dependence towards the long waves (Fig. 6.9(a)). BiI_3 additive to arsenicum tri selenide (2–6 molecular. per cent) shifts the peak of spectral dependence of photosensitivity to the long-wave region (Table 6.2) (Popov, 1997). This photosensitivity increase and the peak shift are accompanied by the rise in dark discharge rate and the decrease in initial surface potential. From the manufacturing point of view, arsenic is not among the favorites due to its high toxicity and relatively high cost.

Photoreceptors on the base of hydrogenated amorphous silicon have the following advantages (Fig. 6.9(b)), Kawamura *et al.*, 1983:

- high photosensitivity throughout visible spectrum and, in the case of multilayer photoreceptors, in the infrared spectrum as well,
- high drum durability ($\geq 10^6$ copies) because of high material micro-hardness (1500–2000 kg/mm^2),
- voltage of any polarity could be used for initial surface charging,
- high temperature stability, allowing the realization of compact machine construction.

However there are at least two technological problems in the manufacturing of a–Si:H photoreceptors. Firstly, it is necessary to substantially increase the film

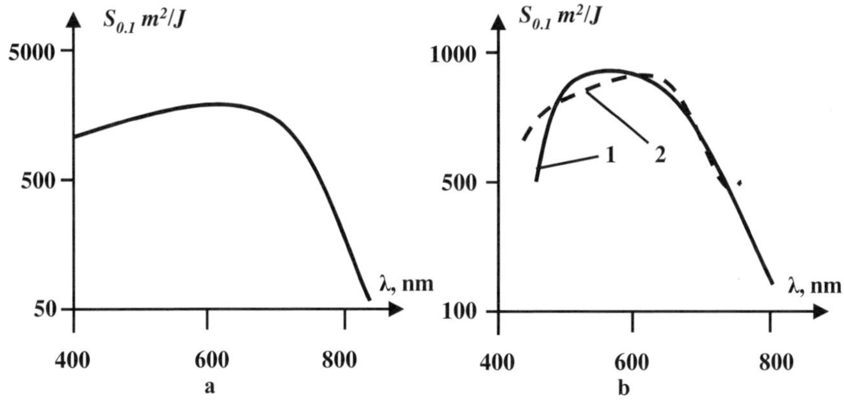

Figure 6.9. Spectral dependences of As_2 Se_3 (a) and a–Si:H (b) photoreceptor sensitivity in the case of negative (1) positive (2) charge.

growth rate as the envisioned layer thickness is 20 μm and more. In the case of glow discharge decomposition method this problem is addressed either by increasing silane concentration (up to 40%) in the gas mixture, or by substitution of silane for polysilanes (Si_2H_6 or Si_3H_8). In the former case, growth rate increases to 5 μm/h. In the latter case, growth rate increases up to 20 μm/h (see Chapter 5). Secondly, it is necessary to increase hydrogenated amorphous silicon resistivity above 10^{13} Ohm/cm, threshold value for electro-photography. Commonly the highest value of resistivity of hydrogenated amorphous silicon films prepared by glow discharge decomposition method does not exceed 10^{11} Ohm/cm (see Chapter 4). The resistivity increase is achieved either by simultaneous boron and oxygen doping of a–Si:H or by creation layers with different type of conductivity in photoreceptor.

The use of sputtering leads to acceptably high values of resistivity in as-prepared films. Use of magnetron sputtering further facilitates growth rate.

Organic photoreceptors. Organic semiconductors are widely used in the production of electro-photographic drums. The spectral characteristics of these materials are very diverse and depend on the type of semiconductor and sensitizing additives. Poly-vinylcarbazole, poly-epoxypropylcarbazole, *N*-isopropyl carbazole and other organic materials are used for photoreceptor build-up. Organic photoreceptor sensitivity is within the limits of 5–50 m^2/J and drum durability is $(5–20)10^3$ copies. These values are an order of magnitude worse than in inorganic photoreceptors. Nevertheless, organic photoreceptors successfully withstand competition through ease of manufacturing (dipping method) and, consequently, low cost.

In contrast with electro-photography, electro-radiography (see Fig. 6.1) image projected onto the semiconductor surface is created by X-rays, and not visible light radiation. This allows to use the method for diagnostics in medicine and in industrial non destructive quality control. The thicker semiconductor layers and additional lead layers on the substrate to facilitate X-ray absorption are characteristic for this application.

Table 6.2 $(As_2\,Se_3)_{100-x}\,(BiI_3)_x$ photoreceptor sensitivity (m^2/J) in infrared range.

	Wave length, nm				
Composition	700	800	850	900	950
$As_2\,Se_3$	150	80	40	10	—
$(As_2\,Se_3)_{98}\,(BiI_3)_2$	260	160	110	70	30
$(As_2\,Se_3)_{96}\,(BiI_3)_4$	300	190	140	95	40
$(As_2\,Se_3)_{94}\,(BiI_3)_6$	240	130	80	40	10

6.1.3 Multi-layers and Variband Photoreceptors

As was mentioned before, the photoreceptor contains several functionally different layers. Therefore tuning the material characteristics to the function of the layer can improve the parameters of the photoreceptor.

In the simplest case, a photoreceptor may contain two layers of semiconductor materials with different gap value. Wide-gap semiconductor is placed near free surface and narrow-gap semiconductor located underneath. In this case the short-wave part of light is absorbed in the wide-gap semiconductor, and lower energy photons pass through this layer and are absorbed in narrow-gap semiconductor. As a result, spectral dependence of photoreceptor sensitivity is expanded. Photoreceptors with double carrier generation layer a-Si_{1-x} N_x–a-Si:H, Se–Se_{1-x} Bi_x and Se–Se_{1-x} Te_x have been created on this basis. Distribution of chemical composition across the thickness of the photoreceptor Se–Se_{1-x} Te_x and its photosensitivity spectral dependence are shown in Fig. 6.10 (Popov, 1997). As one can see from the figure carrier generation region consists of selenium layer with 0.5 μm thickness and Se_{65} Te_{35} layer with thickness of 0.3 μm. Transportation layer is a selenium layer 50 μm thick. Spectral dependence of the photoreceptor sensitivity is superposition of characteristics of the materials used in generation layer (Fig. 6.10(b)).

It is obvious that integral sensitivity of photoreceptor should increase with the number of layers with different gap value in carrier generation region. Natural extension of this process was the creation of variband photoreceptors, in which gap value gradually decreases from the surface to the depth of layer. Dashed line in Fig. 6.10 a shows tellurium concentration changing in this case. Such variband photoreceptors have high value of sensitivity in broad spectral range (from 400 to 800 nm) while maintaining an acceptable level of other electro-photographic characteristics (initial surface potential and dark discharge rate).

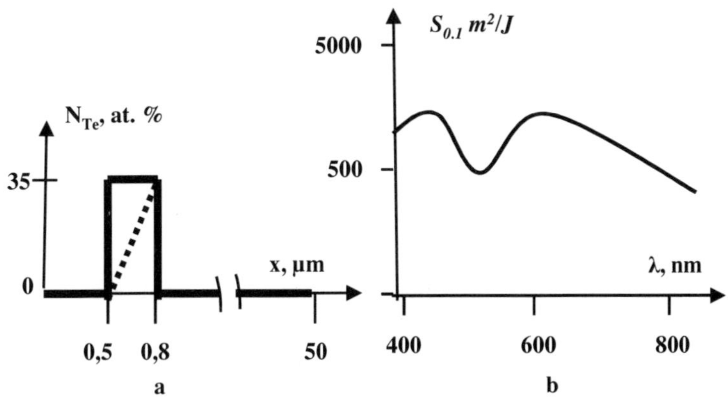

Figure 6.10. Distribution of chemical composition across the thickness of photoreceptor Se – Se_{1-x} Te_x (a), and its photosensitivity spectral dependence (b).

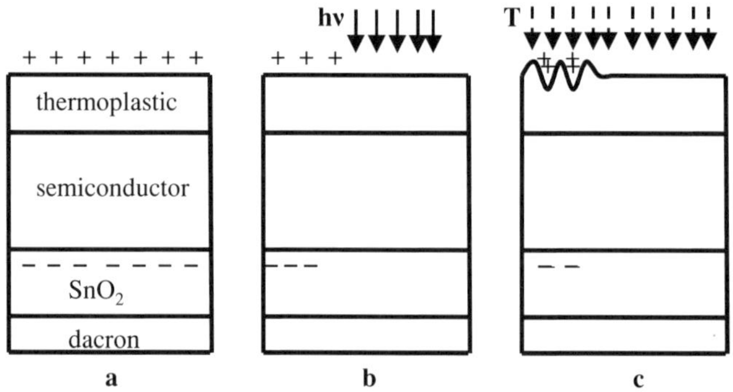

Figure 6.11. Stages of photo-thermoplastic process: charging (a), exposition (b) and development (c).

6.1.4 Photo-thermoplastic Films

Photo-thermoplastic films differ from electro-photographic photoreceptors in that an additional thermoplastic layer is formed on their surface. Thermoplastic is a polymer material that turns to liquid (or flexible) state when heated up to $\sim$ 100 °C and becomes solid when cooled. Thermoplastic layer has high electrical resistivity. Photo-thermoplastic films are prepared on the lavsan (dacron) transparent substrate. The semi-transparent bottom electrode is made of tin oxide or ITO ($In_2O_3 + SnO_2$). High resistance glassy chalcogenide film is used as photosensitive layer ($As_4S_3Se_3$, for example) (Fig. 6.11). Electrical charging (Fig. 6.11(a)) and light exposition (Fig. 6.11(b)) stages are analogous to corresponding stages of electro-photographic process. Electrical field of latent electrostatic image creates an electrostatic pressure which is proportional to the square of surface charge density. Thus, electrostatic pressure distribution replicates the image pattern. Under influence of high temperature thermoplastic turns to flexible state and is deformed under the influence of electrostatic pressure (Fig. 6.11(c)). As a result, the relief corresponding to the latent image appears which is fixed by cooling thermoplastic. This relief alters the phase of readout beam in the process of recorded information reading. Re-heating of photo-thermoplastic film without charging its surface leads to erasing of the information due to the effect of surface tension. Making copies of the recorded information is carried out by hot pressing method.

6.1.5 Vidicons on the Basis of Disordered Semiconductors

Vidicon is television camera pickup tube. Schematic illustration of vidicon design is shown in Fig. 6.12.

Glass substrate 2 with deposited semi-transparent electrode (SnO_2 or SnO_2 $+In_2O_3$)3 and the semiconductor layer 4 are at the bottom of an evacuated glass bulb 8. The gun 6 forms electron beam 5 which can be moved along the layer surface with the help of deflecting plates 7. The image is projected on the semi-

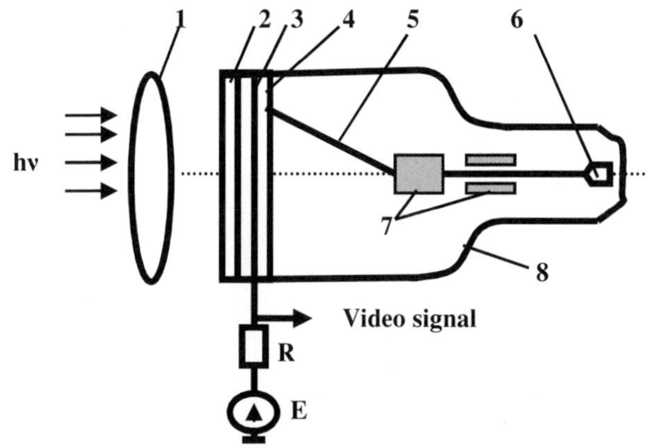

Figure 6.12. Schematic illustration of vidicon design (see text).

conductor layer through an optical system 1, glass substrate and semi-transparent electrode. The surface on the opposite side of the semiconductor layer is scanned by electron beam. The current in the electric circuit — electron beam, semiconductor layer, resistor R and voltage source E depends on semiconductor layer resistance. Resistance of illuminated areas decreases due to photoconductivity and electrical current in the circuit and the voltage drop across the resistor increase. Video signal from the resistor R is fed to an amplifier.

Vidicons based on disordered materials usually make use of As_2Se_3, Sb_2S_3 or As–Se–Te systems. Vidicon sensitivity is determined by the ratio of resistances between lit and unlit layer areas, and hence semiconductor photoconductivity (see Eqn. 3.49). In the case of semiconductor layer thickness of about 1 μm and applied voltage 50 V electric field in semiconductor is about 10^5 V/cm. This corresponds to quantum efficiency close to one (see Fig. 3.30). However, it was mentioned in Sec. 3.4 that quantum efficiency in selenium layers exceeds one if electric field is more than 10^6 V/cm, due to the effect of avalanche multiplication. Moreover, quantum efficiency increases with semiconductor layer thickness at a constant electric field value.

High sensitive and super high sensitive vidicon camera tubes named HARP (High-gain Avalanche Rushing Amorphous Photoconductor) tubes were developed on this basis (Kurashige *et al.*, 1993, Park and Tanioka, 2003). Sensitivity of a tube with layer 2 μm thick increased 10 times compared to 1 μm layer thickness. Layer thickness of 4 μm gave an increase of 40 times and 8 μm of more than 100 times. However, thicker layers require higher target voltage, i. g. 240 V for 2 μm thickness, 460 V for 4 μm and 880 V for 8 μm.

Aerial photographs taken at night with the conventional and HARP vidicons are shown in Fig. 6.13 (Kurashige, unpublished).

High sensitivity of vidicons opens up new possibilities of their application. Let us consider the problem of detecting objects under water at a great depth, as an example. Spectral dependence of sea water optical transmission is shown in

Figure 6.13. Airscape taken at night with the CCD camera (top) and the Super-HARP camera (bottom) (M. Kurashige, unpublished). For color reference, go to page 192.

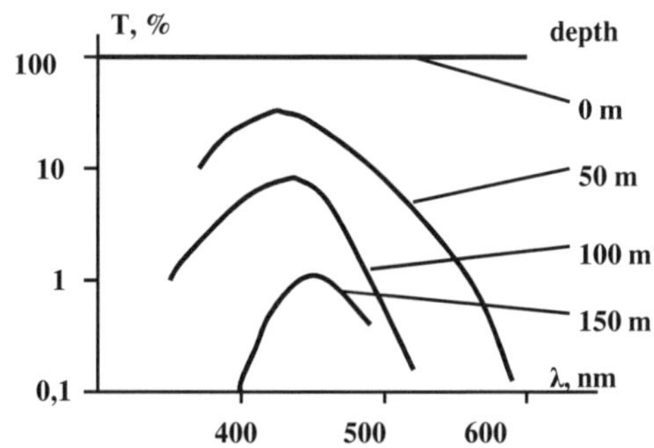

Figure 6.14. Spectral dependence of sea water optical transmission.

Fig. 6.14. Maximum transparency is near 450 nm, which corresponds to the region of maximum photosensitivity of selenium layers. However, the radiation intensity decreases orders of magnitude at a depth of 100–150 meters, even in the area of maximum transparency. On the other hand, the tenfold increase in the HARP tube sensitivity gives ability to detect an object at such a depth, with the same success as conventional vidicon detects it on the water surface (Maruyama, 1992).

Initially, HARP tubes had short operating lifetime due to the low crystallization activation energy of selenium layer and acceleration of crystallization process in high electric field. As a rule, crystallization process begins on either the outside or inside (adjacent to the substrate) surface of the film. A multilayer structure has been used to increase activation energy while maintaining high quantum efficiency. Blocking contacts on both surfaces of selenium film are prepared from materials with higher activation energy: for example, selenium doped with arsenic and arsenicum tri selenide. In this case, the carrier avalanche multiplication occurs only in the pure selenium layer while the areas in contact with the substrate and the electron beam have higher crystallization activation energy due to the presence of threefold coordinated arsenic atoms.

Soon after practical implementation of such structures it was discovered that they do not prevent crystallization in case of operation at lower temperatures. It was found that the device operation below room temperature leads to the appearance of image defects caused by crystallization of the semiconductor layer. This fact is quite surprising, because decreasing temperature, should lead to a decrease in the rate of activation processes in accordance with the Arrhenius Eqn. (1.6). The reasons for crystallization accelerating in the case of device operation at low temperatures have been explained (Resnik *et al.*, 2006) on the basis of photo-structural changes in glassy semiconductors. This offer an opportunity to find a solution to the operating lifetime problem mentioned earlier (see Sec. 6.2).

6.2 DEVICES BASED ON PHOTO-INDUCED TRANSFORMATIONS IN CHALCOGENIDE GLASSES

A number of effects observed in chalcogenide glassy semiconductors are specific only to this class of materials and are not observed in crystals of the same chemical composition. These effects include reversible and irreversible photo-induced changes of optical, chemical and mechanical properties, photo-induced dissolution of metals, photo-crystallization, photo-compression and photo-expansion.

6.2.1 *Photo-structural Changes in Glassy Semiconductors*

Photo-induced structural transformation is based on the following process. Photon absorption leads to electronic excitation, which, in turn, causes changes in the atomic structure of the material. These changes may be both reversible and irreversible. The changes of atomic structure lead to reversible and irreversible transformation of optical, chemical, mechanical and others properties of the material.

Photo-induced transformation of optical properties (absorption coefficient, reflection factor and refractive index) under effect of photons with energy

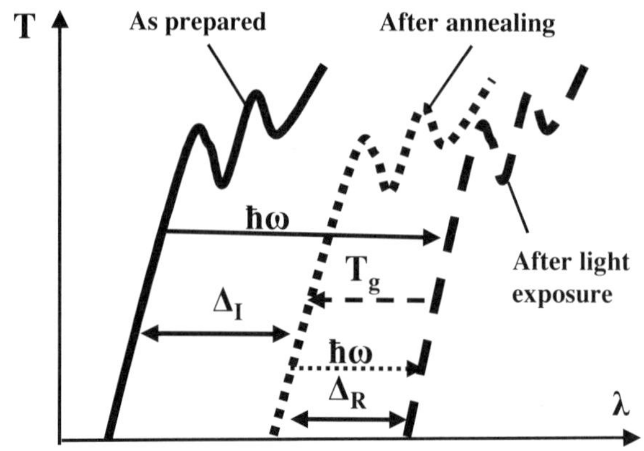

Figure 6.15. Typical changes in optical transmission spectral dependence of chalcogenide film under influence of light radiation ($\hbar\omega$) and annealing (T_g) ($\triangle_R$ is a reversible change and $\triangle_I$ is an irreversible change).

corresponding to the fundamental absorption edge can be observed in various chalcogenide glassy semiconductors: two-component glasses of As–S, As–Se, Ge–S, Ge–Sesystems, three-component systems (As–S–I, As–S–Se, As–Se–Ge, As–Se–Cu) and others. Most of the research is done on thin films (thickness less than 1 μm) but similar effects are observed in bulk samples (Zakery and Elliott, 2003).

Typical changes in optical transmission spectra of chalcogenide film under influence of light radiation and annealing are shown in Fig. 6.15.

Irradiation of as-prepared film by photons with energies corresponding to the fundamental absorption edge of the material shifts the optical transmission spectrum to the long-wave region. It corresponds to photo-darkening of the film. The annealing at the temperature near glass transition value (T_g) leads to curve shift in the opposite direction to some intermediate position. The subsequent cycles of irradiation and annealing are accompanied by reversible shifts of the curve between the last two positions. Thus, photo-induced changes of optical properties of chalcogenide glassy films have both reversible part (irradiation of annealed film) and irreversible part (first irradiation of as-prepared film). The values of these changes depend on the film's chemical composition, methods and conditions of film preparation, photon energy, temperature and other factors.

Dependence of the reversible part of arsenicum tri sulphide transmission spectrum shift on irradiation photon energy is shown in Fig. 6.16 (Tanaka, 1980). Data is normalized to the number of absorbed photons. Spectral dependence of material optical absorption coefficient is shown in the figure as well. It was mentioned in Sec. 3.3 that value of optical gap may be determined as photon energy corresponding to optical absorption coefficient $\alpha = 10^3 \mathrm{cm}^{-1}$. One can see from the figure, that maximum shift of the optical transmission spectrum is observed at photon energy approximately equal to the gap value (2.45 eV). The spectrum shift remains unchanged with increasing photon energy. In contrast, the

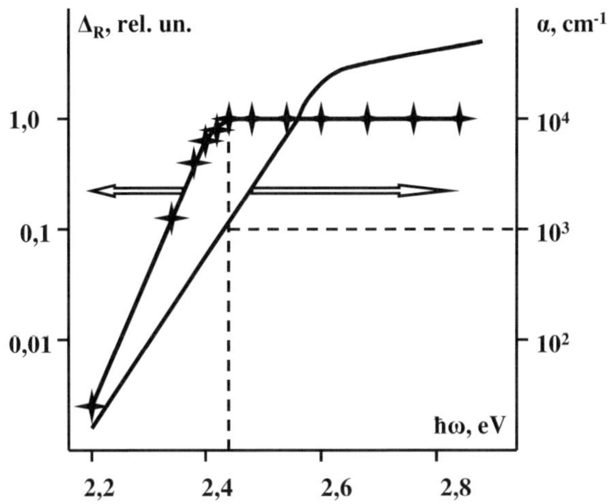

Figure 6.16. Dependence of reversible part transmission spectrum shift (Δ_R) on irradiation photon energy and spectral dependence of absorption coefficient (α) of As_2S_3 film.

decrease of the photon energy leads to a sharp reduction of effect value by orders of magnitude.

Photo-induced changes in optical properties of thermally evaporated films from the systems As–Se with various component ratios are shown in Table 6.3 (Gurevich *et al.*, 1976).The films were exposed to radiation with a wavelength 633 nm at a temperature of 300 K. The value of reversible and irreversible photo-induced property changes grows with increasing arsenic content. At the same time, the material's sensitivity to photo-induced changes decreases.

The value of photo-induced changes in optical properties depends on the sample temperature during irradiation. Moreover, under certain conditions, irradiation may cause an increase in the material's optical transmission (photo-bleaching). Reversible part of As_3Se_2 film optical transmission changes under radiation of helium-neon laser at different temperatures are shown in the Fig. 6.17 (Averynov *et al.*, 1980, Kolomiets, 1981). Transmission changes from point *A* to point *B* correspond to the photo-darkening of the film irradiated at room temperature. Heating of the irradiated film leads to a slight decrease in transmission firstly due to the temperature dependence of gap value (section *BC*). Sharp increase in transmission occurs when the temperature reaches the annealing temperature value (section *CD*). Section *DA* corresponds to transmission changes of annealed film with temperature decreasing. If irradiated at room temperature film is heated to temperature T_1 and then irradiated again, the transmission increases from point *E* to point *F*. In other words, photo-bleaching of the film takes place. On the other hand, irradiation at the same temperature of annealed film (point *G*) reduces the optical transmission to the same value, characterized by the point *F*. Finally, if irradiated at room temperature film is heated while being irradiated its transmittance will change along the curve *BLFD*. Thus, the final value of the film

Table 6.3 Photo-induced changes in optical properties of films from the systems As–Se ($\lambda = 633$ nm, $T = 300$ K).

Composition	Irreversible changes			Reversible changes	$T_{annealing}$, K	T_g, K
	Δ_I, eV	contrast	refractive index changes	refractive index changes		
Se	-	-	-	-	-	~305
As_2Se_7	-	1,5	$\leq$ 0,02	-	370-400	389±2
As_2Se_5	0,025	2,0	$\leq$ 0,02	-	408-418	408±2
As_2Se_3	0,06	8,0	0,13	0,02	463-473	464±2
AsSe	0,145	80-100	0,21	0,13	463-468	455±2
As_3Se_2	0,19	180-200	0,30	0,24	423-438	414±2

Δ_I is irreversible shift of optical transmission spectrum.

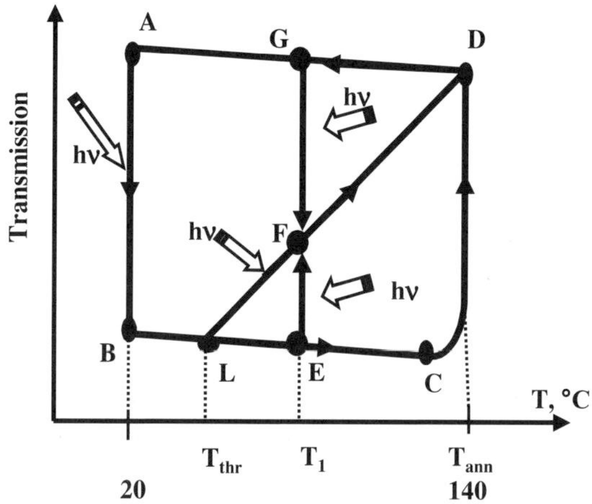

Figure 6.17. Reversible changes of optical transmission in As_3Se_2 film at various temperatures.

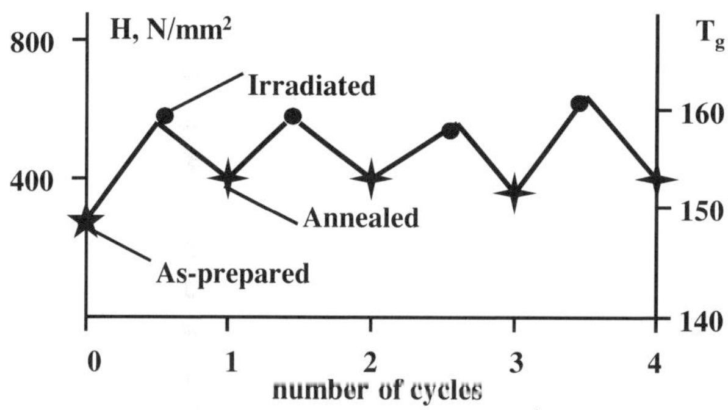

Figure 6.18. Changes in the micro hardness and glass transition temperature in As_3Se_2 film under irradiation and annealing.

optical transmission is determined only by the temperature of the last exposure and does not depend on the sequence of irradiation and annealing processes.

Photo-induced anisotropy of optical properties induced by linearly polarized light is also observed in glassy chalcogenide films (Zakery and Elliott, 2003).

Along with the optical properties photo-induced changes in the electrical and photoelectrical properties, micro hardness, glass transition temperature and the rate of material etching are observed in glassy chalcogenide films. These changes have reversible and irreversible components as well. The changes in the micro hardness and glass transition temperature of As_3Se_2 films under the influence of light exposure and annealing are shown in Fig. 6.18 (Gurevich *et al.*, 1976b). Characteristic values increase under irradiation and decrease with subsequent annealing. The relative changes of micro hardness reach 40–50%. Density of the films varies similarly. The increase of the density of the material leads to a decrease of film thickness (photo-contraction effect).

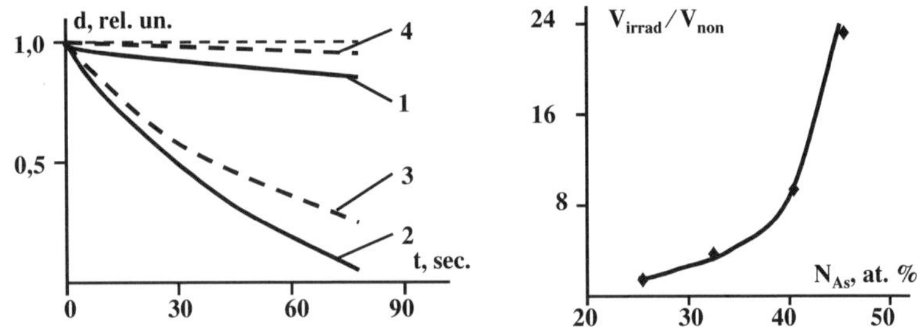

a b

Figure 6.19. Thickness of non-irradiated (curves 1 and 3) and irradiated (curves 2 and 4) areas of AsSe film dependencies on etching time in NaOH (curves 1 and 2) and $K_2Cr_2O_7$ (curves 3 and 4) (a) and dependence of dissolution rates ratio of irradiated (V_{irrad}) and non-irradiated (V_{non}) areas of As–*S* system films on arsenic concentration (b).

The effect of photo-stimulated changes on the etching rate of chalcogenide glasses in certain solvents is very interesting from the application point of view. Alkaline enchants dissolve the irradiated areas several times faster than non-irradiated areas (Fig. 6.19(a)) (Kolomiets, Lyubin and Shilo, 1978). Therefore, after processing of exposed film in an alkaline etchant, positive image is obtained. However, if the same film is processed in potassium dichromate ($K_2Cr_2\ O_7$), the effect will be opposite. Irradiated areas of the film are practically insoluble, and non-irradiated areas dissolve quickly and the negative image is obtained as a result. Thus, both positive and negative image can be obtained on the same material, depending on the type of solvent.

Photo-induced changes in the etching rate of chalcogenide glasses accelerate with increasing arsenic concentration. The dissolution rates ratio of irradiated and non-irradiated areas of As–S system films increases by 12 times with arsenic concentration increasing from 25 to 45 at.% (Fig. 6.19(b)) (Petkov and Fedorov, 1982).

Changes of chalcogenide glass properties occur not only under light irradiation but also under the action of electrons and X-ray. These effects have reversible and irreversible components and are associated with structural changes in the material. Blackening curves of glassy $As_{65}\ Se_{35}$ film electron diffraction patterns obtained after different periods of electron exposure are shown in Fig. 6.20. As seen from the figure, changes in the diffraction pattern occur in the area from 1.5–4.5 Å^{-1} during the first two minutes of sample irradiation. The intensity of the second and, especially, the third peaks decreases and their mutual rapprochement takes place. At the same time, there's no significant changes in height and position of the first peak at 1.1–1.2 Å^{-1}. After approximately 2 minutes (at 50 kV accelerating voltage) diffraction pattern is stabilized and further exposure of the sample by electrons does not lead to changes in it.

These results are a direct evidence of structural changes occurring in the material under electron radiation. The observed broadening of the peaks in

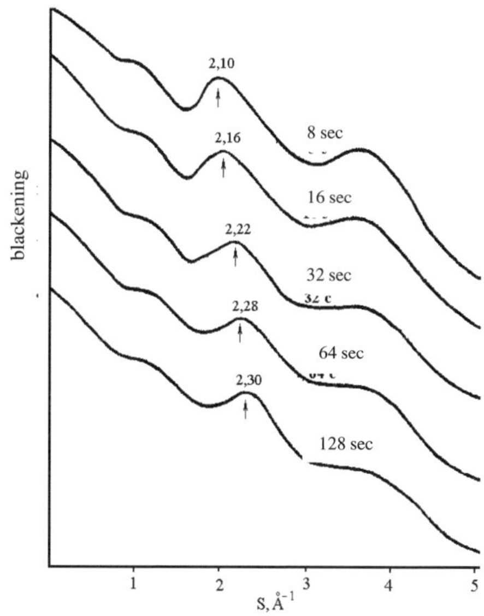

Figure 6.20. Blackening curves of glassy As_{65} Se_{35} film electron diffraction patterns obtained after different periods of electron exposure (accelerating voltage is 50 kV).

the diffraction pattern indicates increasing of structural disorder under electron radiation. It should be noted that it is related to changes in the short-range order of atomic arrangement, since the peak at $S = 1.1 - 1.2$ Å^{-1}, which characterizes the medium-range order, remains practically unchanged.

Analysis of experimental results shows that photo-induced property changes described above are observed in materials that meet at least two conditions: chalcogen atoms have to be in the material composition and the material must be in non-crystalline state. Initially, the photo-induced changes of chalcogenide glass properties were associated with photo-chemical reactions, accompanied by formation of individual component clusters of the material. However diffraction and other studies have established that these changes were caused by changes in the atomic structure (Tanaka, 1981). A number of phenomenological models have been created to explain the photo-induced changes in the properties of chalcogenide glassy semiconductors (Kolobov and Shimakawa, 1996). These models are based on presence of atoms in the structural network that can be in two different metastable states (Fig. 6.21). The state with lower energy is a ground, but a state with higher energy is excited. Optical and thermal transitions between these two states are taking place. Energy versus configuration dependencies have parabolic form. It is assumed that the parabolas, corresponding to two states, have significantly different steepness (Fig. 6.22) (Kolomiets, 1981).

As seen from Fig. 6.22, the energy required for the optical transition from ground state 1 to excited state 2 (direct transition $I_{\text{opt}1\to2}$ by the selection rule) is significantly less than the energy of optical transition from state 2 to state 1 ($I_{\text{opt}2\to1}$). As a result, the first process is dominating under irradiation, and most

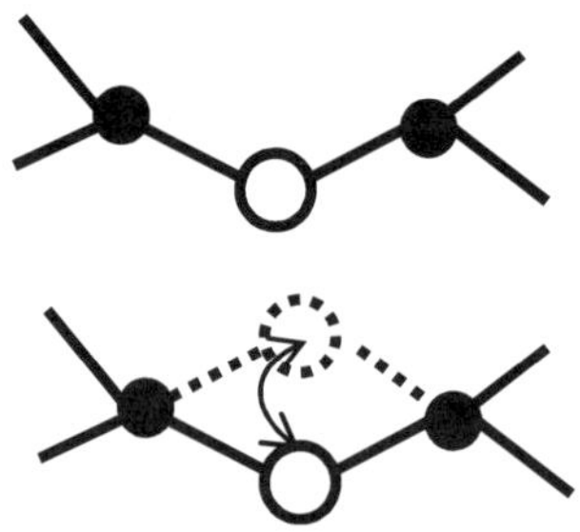

Figure 6.21. Schematic representation of the bistable state (open circles denote chalcogen atoms, dark circles denote pnictide atoms).

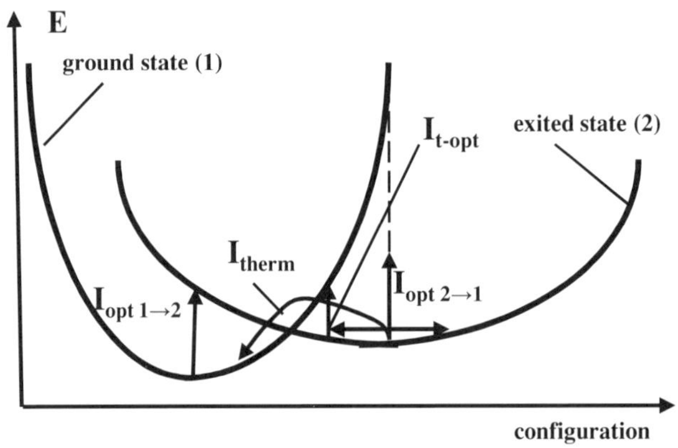

Figure 6.22. Illustration of photo-induced changes on the base of configuration diagram (I_{opt} is optical transition, I_{term} is thermal transition and I_t − opt is thermo-optical transition).

of bistable sites are transferred to state 2. This leads to photo-darkening and to changes in other properties described above. Recurrence of atoms to state 1 is possible by thermal transition I_{therm} during annealing at temperatures close to glass transition temperature. However, there is another possibility of transferring bistable sites from state 2 to state 1. Heating of the sample reduces the energy required for the optical transition from state 2 to state 1, due to increasing thermal vibrations of atoms. As a result bistable site is transferred to state 1 by means of thermo-optical transition I_{t-opt}. This explains the effect of photo-bleaching under material irradiation at heightened temperature (transition from point E to point F in Fig. 6.17).

The effect of semiconductor layer crystallization accelerating in the case of vidicon operation at low temperatures (see Sec. 6.1.5) can be explained based on the above model. Let us consider the configuration diagram of glassy and crystalline states of the semiconductor (Fig. 6.23). Left parabola in the figure corresponds to the glassy state, and the right parabola corresponds to the crystalline state. Activation energy of crystallization is equal to barrier height E_{cryst}. Optical transitions E_{opt} occur under irradiation and excited (defect) states corresponding

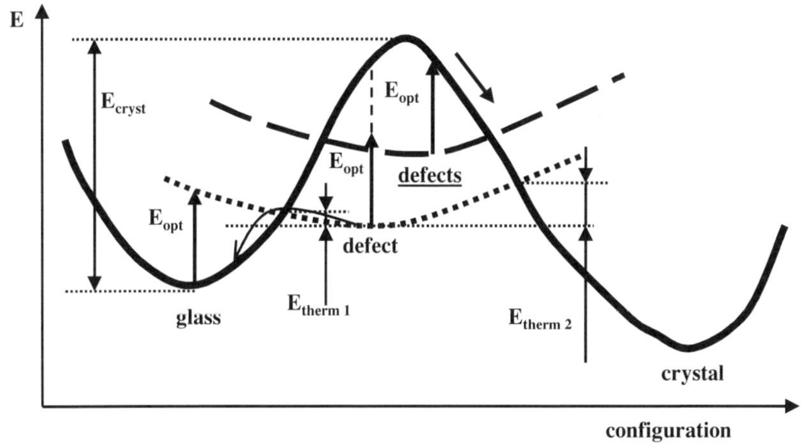

Figure 6.23. Configuration diagram illustrating the effect of light-induced selenium layer crystallization at low temperature (see text).

to the dotted parabola are formed. Reverse optical transition to the ground glassy state is not possible because of different steepness of the parabolas. Energy required for thermal transition to crystalline state is equal to E_{therm2} and energy required for thermal annealing to glassy state is equal to $E_{therm1} < E_{therm2}$. However there is not enough energy for any of these transitions at low temperatures and photo-induced excited states are accumulated in the material. This leads to energy increase and to a shift of the parabola corresponding to the excited state (the dashed curve in the figure). At a certain moment the photon energy becomes sufficient for optical transitions from the excited states. But, as can be seen from the figure, these transitions lead to material crystallization.

The above configuration diagram not only explains the crystallization acceleration in the case of vidicon operation at low temperatures, but also shows the way to solve the problem. Elevating the temperature of the selenium vidicon target to 35°C during illumination acts to instantaneously anneal the photo-induced excited defect states and eliminates the crystallization (Resnik *et al.*, 2006).

The common advantage of photo-induced transformation based devices is high resolution (at least 10^3 lines per millimeter), and the general drawback is the low photosensitivity (10^{-2}–10^1 cm^2/J). Other advantages of these media are the possibility of multiple recording and high diffraction efficiency in hologram recording, due to significant changes in the refractive index. Photo-induced changes in the etching rate make it possible to obtain relief images. It also provides high diffraction efficiency in holographic recording and the possibility of material application as a submicron photoresists in microelectronics. Existence of electron-stimulated and X-ray stimulated changes of chemical properties makes it possible to use these materials in electronic lithography and X-ray lithography.

Low propagation losses and photo-induced effects make chalcogenide glasses suitable candidates for the fabrication of devices such as gratings and waveguides

(Zakery and Elliott, 2003). Transmissive diffraction gratings for use at infra-red wavelengths have been fabricated in chalcogenide glasses using the photo-darkening effect. These gratings can be used as efficient beam combiners, couplers and have significant application in monochromators, laser-tuning devices, shapers, coupling of optical fibers. Planar channel waveguides with very low losses (< 0.3 dB/cm) have been fabricated in different compositions of chalcogenide glasses using the photo-darkening effect in thermally evaporated, sputtered and pulsed-laser deposited films.

6.2.2 Photo-induced Metal Dissolution in Chalcogenide Glasses

It has been shown (Kostyshin *et al.*, 1966) that the diffusion of some metals (Ag, Cu) in chalcogenide glasses As–S, As–Se and some other systems is accelerated by orders of magnitude under illumination by light of greater than band edge energy. This process readily occurs at room temperature at illumination level as low as 10^{-4} J/cm^2. Until now the nature of this effect has only been explained at qualitative level. The photons pass through a semi-transparent silver film of about 30 nm thick and are absorbed in chalcogenide film (Fig. 6.24). As a result, the electron — hole pairs are generated in chalcogenide film. Since the mobility of electrons in these materials is significantly less than the hole mobility, the electrons are captured by traps and charged C_1^- centers are formed. The holes diffuse from the generation region and recombine with the electrons of silver atoms. The diffusion of positively charged silver ions is accelerated by the Coulomb interaction with the negatively charged C_1^- centers.

Photo-doping leads to changes in optical and chemical properties. It has been shown that chalcogenide films of Ge–S and Ge–Se systems photo-doped with silver are extremely resistant to etchants used in semiconductor manufacturing. This, together with high resolution (see Table 6.1), allows to create photo-resists and electron-resists for high and super high resolution (Masters and Goldberg, 1981).

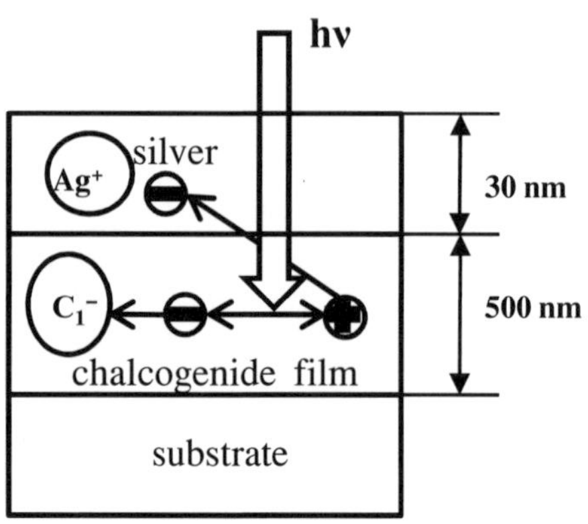

Figure 6.24. Schematic illustration of photo-doping process (see text).

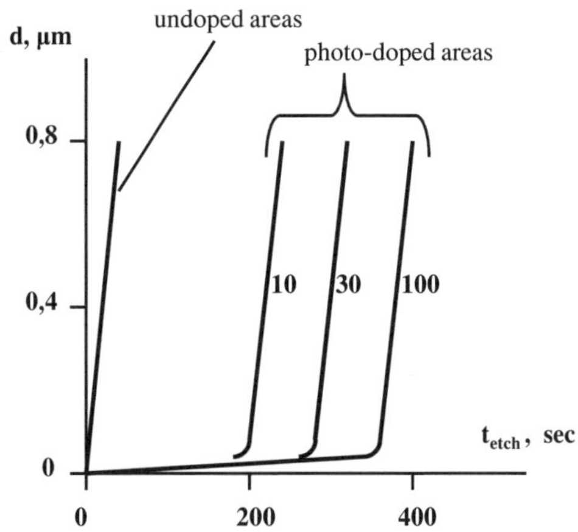

Figure 6.25. Dependences of the plasma etching depth of Ge_{25} Se_{75} photo-resist on etching time in CF_4 (numbers on the curves indicate the exposure time, sec).

Photolithographic process based on photo-induced silver dissolution is as follows. Chalcogenide film Ge – Se system with thickness of less than 1 μm and a thin silver film (30–50 nm thick) are deposited on the surface of oxidized silicon wafer sequentially. This structure is exposed. Silver on the exposed areas diffuses into chalcogenade film. Silver remaining on the unexposed areas is removed in acid solution. Then wet or plasma etching of chalcogenide film and SiO_2 is carried out. The depth of the plasma etching of Ge_{25} Se_{75} photo-resist dependencies on etching time in CF_4 are shown in Fig. 6.25 (Mizushima and Yoshikawa, 1982). These results indicate high efficiency of photolithographic process. Transmission electron microscope study shows that the spatial resolution for photo-doping effect is at least as fine as 10 Å.

The following advantages of these resists can be mentioned:

- high resolution,
- high resistance to etchants which allows to use thin resist films and, as a consequence, increase the resolution of the photolithographic process overall,
- high hardness , which reduces the probability of defect formation in the manufacturing process,
- high atomic mass of elements in the resist composition effectively protects wafer surface in the case of ion implantation doping.

6.2.3 Photo-induced Phase Transitions

The ability to store information is an important aspect of our civilization's development. Among multiple methods of information storage, the optical data storage based on the first-kind phase transitions (evaporation, melting, crystallization and amorphization) in chalcogenide films is one of the most successful.

Photo-induced evaporation leads to formation of surface relief (pits) and to changes in optical reflection from exposed areas. Optical contrast due to recoded pits is determined as follows:

$$S = (I_{R1} - I_{R2})/(I_{R1} + I_{R2}) \tag{6.3}$$

where I_{R1} and I_{R2} are light intensities reflected from original surface and pit area respectively. This kind of optical data storage is named recordable or write-once-read-many-times memory (Moller *et al.*, 2003).

Optical data storage based on crystalline-amorphous transitions uses significant property differences of crystalline and amorphous phases. At the same chemical composition band gap of amorphous phase is 30–50 % larger than crystal band gap, dielectric constant is 2–2.5 times and resistivity is 10^4–10^5 times larger. The reversibility of the crystallization-amorphization process allows fabrication of rewritable memory (write-and-read-many-times memory). Phase transitions between amorphous and crystalline states originally have been applied to creation of non-volatile memory switchers on the base of multi-component chalcogenide films (Ovshinsky, 1968). The difference in electrical conductivity of amorphous and crystalline phases has been used in these devices (see Sec. 7.2.2).

Optical data storage or optical disc is a flat circular polymer substrate (usually polycarbonate) about 1.2 mm thick with deposited recording (phase-change material), reflective and protective layers. The encoding pattern follows a continuous spiral path covering the entire disc surface and extending from the innermost track to the outermost track. The distance between centers of two neighboring tracks lies in the range 1.6–0.32 µm.

The concept of phase-change optical recording can be described as follows (Fig. 6.26). Short high intensive laser pulse locally heats the crystalline phase of phase-change material above melting temperature (transition 1). Melted volume quenches into amorphous phase after pulse finishing (transition 2) due to high

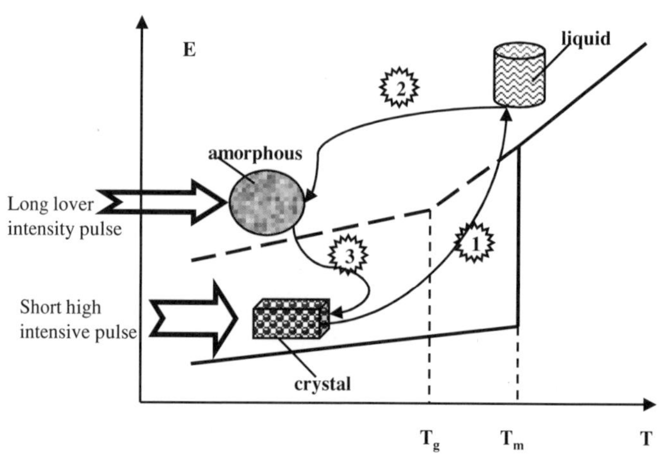

Figure 6.26. Schematic illustration of phase-change optical recording (see text).

cooling rate (10^8–10^9 K/sec). This amorphous dot has lower optical reflectivity in comparison with surrounding crystalline phase and can be easily detected with low intensity laser beam. To erase the stored information laser pulse with lower intensity and longer duration locally heats amorphous dot to a temperature slightly above the glass transition temperature. It leads to increasing mobility of atoms and crystallization of amorphous dot (transition 3).

Phase-change material in optical disc should satisfy the following requirements (Wuttig and Yamada, 2007):

- high speed of phase transition induced by nanosecond laser pulse,
- high optical contrast determined by considerable difference in refractive index of amorphous and crystalline phases,
- long thermal stability of amorphous state to ensure data retention (at least several decades at room temperature),
- large cycle number of amorphous-crystalline state transitions without material decomposition (more than 10^5 cycles),
- high chemical stability and water resistivity.

The above requirements are rather strict and contradictory. For example, amorphous dots should be stable at room temperature at least 10 years (10^8 sec). On the other hand fast crystallization within several tens of nanosecond (10^{-8} sec) must occur under laser pulse at the temperature range between glass-transition and melting temperatures. It means that increasing temperature by a few hundred degrees has to change phase-transition time by 16 orders of magnitude. Therefore very few materials have the properties' combination mentioned above. As a result of development process the following systems and materials (Ge–Sb, Ge–Sb–Te, $Ge_2\,Sb_2\,Te_5$ named GST 225, AgInSbTe named AIST) were found to be the best in terms of speed and stability.

It should be noted that fast and stable phase-change transitions in these materials are difficult to quantitatively describe by means of conventional melting, amorphization and crystallization processes. Based on extended X-ray absorption fine structure spectroscopy (EXAFS, see Sec. 2.3) of $Ge_2\,Sb_2\,Te_5$ Kolobov *et al.*, 2004 concluded that phase changes under laser pulse occur without material melting in a conventional sense. Germanium atoms occupy the octahedral position in the crystalline state and have stronger covalent and weaker bonds (Fig. 6.27(a)). An intense laser pulse induces rupture of the weaker bonds due to electronic excitation and germanium atom flips into tetrahedral position corresponding to the amorphous state ("umbrella-flip" structural transformation, Fig. 6.27(b). Notice that three covalent bonds remain intact. This structural transformation involves a change in electron orbital hybridization and accounts for a large change in the optical properties of the material in phase transition.

Data storage capacity of optical discs is directly related to the size of amorphous/crystalline dots. Dot size in the first place depends on wavelength of laser used for recording. That is why decreasing the recording laser wavelength has become the main way to increase overall density of information storage achievable

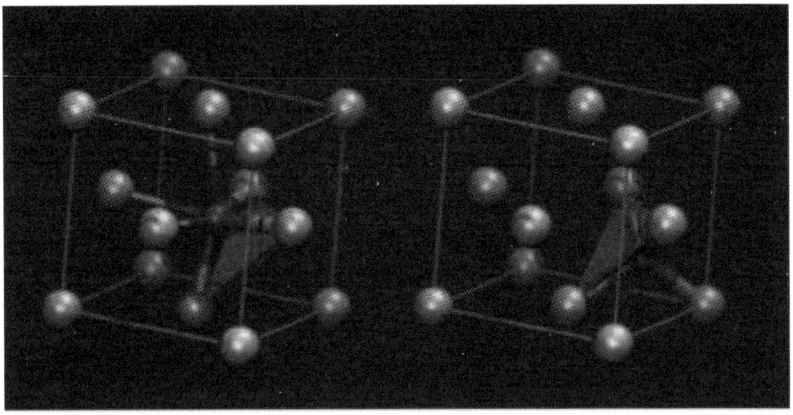

a b

Figure 6.27. Fragments of the local structure of GST225 around germanium atoms in the crystalline (a) and amorphous (b) states (stronger covalent bonds are shown as thicker lines whereas weaker bonds are shown as thinner lines) (Kolobov *et al.*, 2004). For color reference, go to page 193.

Table 6.4 Changing in refractive index (n) and absorption coefficient (k) in Ge–Sb–Te alloys for a wavelength 405 nm.

Composition	Ge content, at. precent	$\Delta n = n_{cryst} - n_{amorph}$	$\Delta k = k_{cryst} - k_{amorph}$
$Ge_2Sb_2Te_5$	22	1.20	1.05
$Ge_4Sb_2Te_7$	31	1.26	1.09
$Ge_6Sb_2Te_9$	35	1.45	1.27
$Ge_8Sb_2Te_{11}$	38	1.42	1.35

in a single disc. However the problem of good optical contrast has appeared at reduced wavelengths. The GST225 alloy that showed good optical contrast in the infrared and red spectral range is not ideal for shorter wavelengths. Optical property changes due to amorphous-crystalline state transitions in alloys of the system Ge–Sb–Te with different chemical composition are presented in Table 6.4 for wavelength of 405 nm. As the table shows, the difference in the optical properties and, consequently, the optical contrast increase with increasing germanium concentration (Yamada, 2002). Alloys such as $Ge_8Sb_2Te_{11}$ or those with more germanium content possess higher optical contrast and are preferable for short wave optical recording.

Along with decreasing wavelength of the recording laser, development of optical memory went towards improving the optical and mechanical systems of disc drive and software algorithms. Progress of the phase-change optical discs is shown in Table 6.5 and Fig. 6.28 (Wuttig and Yamada, 2007). Reduction of the laser

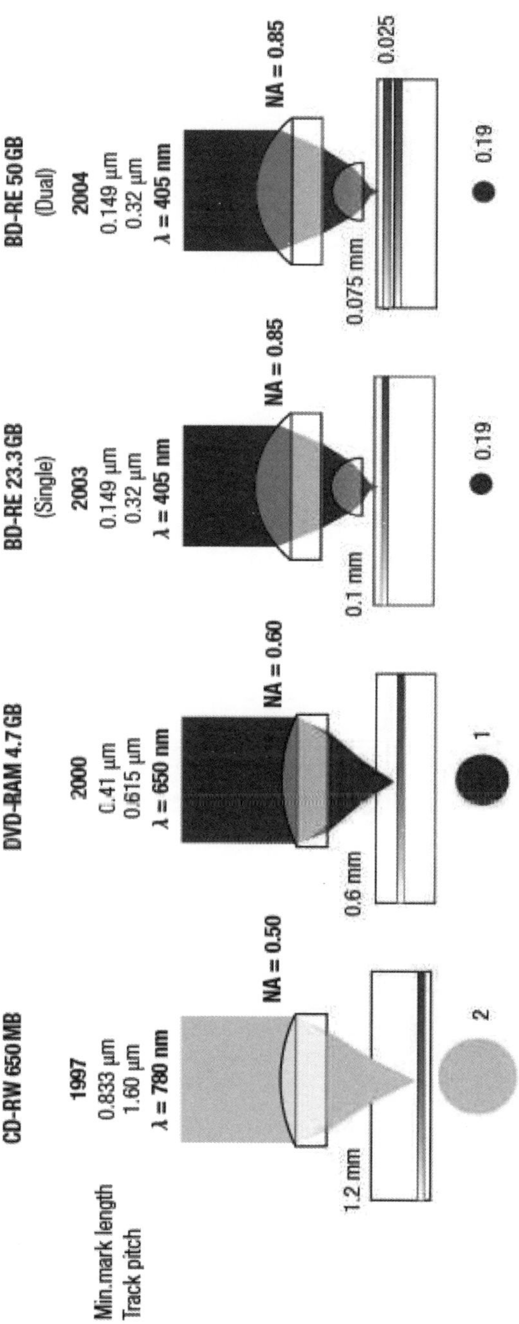

Figure 6.28. Comparison of Rewritable optical disc charactristics. For color reference, go to page 194.

Table 6.5 Comparison of rewritable optical disc characteristics.

Year	1997	2000	2003 (single)	2004 (dual)
Recording capacity	650 MB	4.7 GB	23.3 GB	50 GB
Laser wavelength	780 nm	650 nm	405 nm	405 nm
Track pitch	1.60 μm	0.615 μm	0.32 μm	0.32 μm
Numerical aperture	0.50	0.60	0.85	0.85
Minimal mark length	0.833 μm	0.41 μm	0.149 μm	0.149 μm
Distance from disc surface to phase-change layer	1,2 mm	0.6 mm	0.1 mm	0.075 + 0.025 mm
Cross-section of focused beam	2 μm	1 μm	0.19 μm	0.19 μm

wavelength from infrared to red and after that from red to blue-violet, increase of the numerical aperture of the lens from 0.5 to 0.85 and doubling of the recording layers from single to dual ensured the rise of recording density by two orders of magnitude and increase of the data-transfer rate by three orders of magnitude.

Chapter Seven

Photoelectric and Electronic Devices Based on Disordered Semiconductors

7.1 PHOTOVOLTAIC DEVICES

Photovoltaic cells, or solar cells, transform light (usually sunlight) directly into electric current. Direct conversion of solar energy into electricity is very attractive for several reasons:

- sunlight is free and inexhaustible,
- sunlight is abundant (power of solar radiation at Earth's surface is 10^{16} TW, which is 4 orders of magnitude greater than the need for mankind in the production of energy today),
- photovoltaic systems burn no fuel and thus produce neither atmospheric emissions (e.g. greenhouse gases) nor hazardous waste to cause detrimental effects on the environment,
- since photovoltaic systems have no moving parts, they are silent and have high reliability.

The first solar cell based on single-crystal silicon has been manufactured in 1954 and had efficiency of about 6%. Significant progress in the development of solar cells based on single-crystal silicon has been made in the 1960s fuelled by the development of space technology. At present, efficiency of such photovoltaic cells is more than 24% (the theoretical maximum value is 29%). Solar cells based on single crystals of gallium arsenide were also created in the same period. However, production cost of solar cells based on single crystals is rather high. This leads to the high cost of electricity produced. Cost reduction by a factor of 10 is necessary for solar cells to be competitive with the traditional sources of electricity. The challenge is great, considering high cost of single crystals themselves. The solution lay with the development and manufacture of thin-film solar cells based on polycrystalline or amorphous semiconductor material. While such cells have lower efficiency, the significantly lower production costs and possibility of manufacturing elements with a large surface area make it possible to reduce the cost of produced electricity.

Disordered Semiconductors: Physics and Applications by Anatoly Popov

www.panstanford.com
978-981-4241-76-2

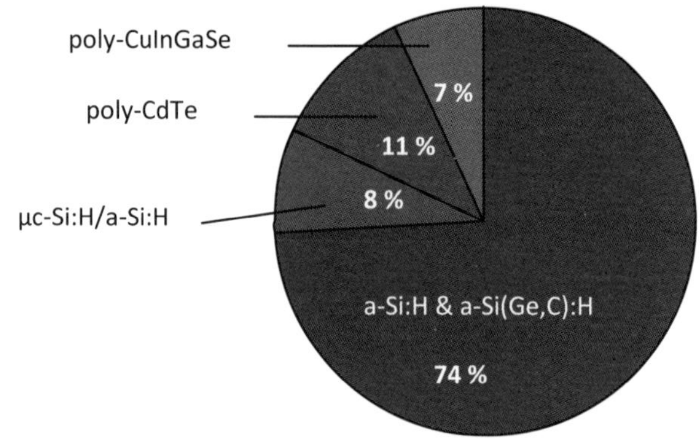

Figure 7.1. World production (sales) of thin-film solar sells. For color reference, go to page 195.

Thus, today there are two trends in photovoltaic solar cell development:

- solar cells based on single crystal characterized by high efficiency and high cost,
- solar cells based on thin film with non crystalline or even organic structure characterized by lower efficiency and significantly lower cost.

Global annual solar cell production has been growing at about 36% per year for the previous decade and production levels exceeded 1 Gigawatts in 2004, 1.4 Gigawatts in 2005 and 7.3 Gigawatts in 2008. World production (sales) of thin-film solar cells based on various semiconductor materials is shown in Fig. 7.1. As can be seen from the figure, solar cells based on hydrogenated amorphous silicon, its alloys and microcrystalline silicon represent more than 80% of the total production of thin-film photovoltaic cells.

It should be noted that the optical absorption of sunlight in the hydrogenated amorphous silicon is 20 times higher than in single-crystal silicon. This allows using hydrogenated amorphous silicon films with thickness of up to 1 micron in solar cells instead of expensive single-crystal silicon wafers with thickness of hundreds of microns.

The principle of photovoltaic cells is based on the internal photoelectric effect (see Sec. 3.4). Irradiation with light in the fundamental absorption region of the semiconductor structure with a *p-n* junction leads to the generation of electron-hole pairs. Concentration of electron-hole pairs in the generation region increases, so that they diffuse to the *p-n* junction. The electric field of *p-n* junction separates electrons and holes (Fig. 7.2(a)). As a result, the open circuit voltage (U_{OC}) across the terminals of photovoltaic cell appears if external electric current equal zero. The magnitude of this voltage is determined by *p-n* junction contact potential difference (barrier height). If the external circuit is closed (load resistance is equal to zero), the current in the circuit will be equal to short-circuit current (I_{SC}), determined by the internal resistance of the cell. In the case of finite values of load resistance, current and voltage will have some intermediate values. Current-

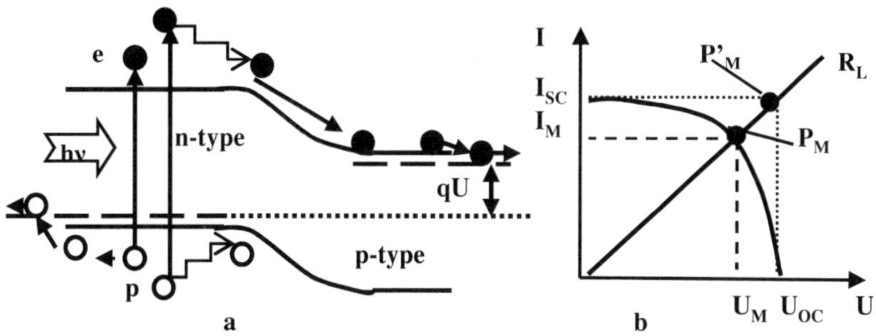

Figure 7.2. Schematic illustration of photovoltaic cell principle (a) and current-voltage characteristic (b).

voltage characteristic of photovoltaic cell is shown in Fig. 7.2(b).

The efficiency of a photovoltaic cell can be written as follows

$$\eta = I_M U_M / P_i = (I_{SC} U_{OC} FF)/P_i \tag{7.1}$$

where I_M and U_M are current and voltage of photovoltaic cells at the point of maximum power, *FF* is fill factor and P_i is power of solar flux incident on the cell surface under standard test conditions (temperature 25°C and an irradiance of 100 mW/cm^2 with an air mass 1.5 (AM1.5) spectrum). AM1.5 corresponds to the spectrum of sunlight incident on a clear day with the sun at an angle of 48° above the horizon. Fill factor is determined from the expression (7.1) as follows:

$$FF = (I_M U_M)/(I_{SC} U_{OC}) \tag{7.2}$$

and is the ratio of maximum output power of photovoltaic cell (point P_M in the Fig. 7.2(b) to the power of the cell with an idealized rectangular current-voltage characteristic (point P'_M in the Fig. 7.2(b)).

Conditions for efficient conversion of solar energy into electricity in photovoltaic cell are as follows:

- significant part of solar radiation has to be absorbed in the active region of the cell,
- photo-generated electrons and holes must be effectively collected on the contact electrodes,
- potential barrier of *p-n* junction should be large enough, since it determines the value of the cell's open circuit voltage,
- internal resistance of the cell, which includes the resistances of semiconductor layers, contacts etc., should be small enough, since it determines the value of short-circuit current.

The efficiency of solar radiation absorption is determined by the absorption coefficient of semiconductor material in the range of photon energies from 1.0 to 3.5 eV and by the thickness of the semiconductor. At the same time, the thickness of the semiconductor layer affects the collection efficiency of photo-generated electrons and holes, and should not exceed the diffusion length of carriers. Otherwise, the charge carriers recombine before their separation by electric field of *p-n* junction. In the case of single-crystal semiconductors, the above requirements are consistent with a large diffusion length of carriers. In disordered semiconductors, due to lack of long-range order in the arrangement of atoms and high density of localized states, the diffusion length of carriers is small. Therefore, these requirements are in conflict and require special solutions for implementation.

For separation and subsequent collection of photo-generated charge carriers in the semiconductor layer an internal electrical field has to be created. In the cells based on single-crystal semiconductors this field is created by the space charge of *p-n* junction.

The width of *p-n* junction space charge in non-crystalline semiconductors is small because of the high density of localized states. Consequently, separation of charge carriers occurs only in a small part of the semiconductor layer. This situation, together with a small diffusion length of charge carriers leads to low collection efficiency of carriers. To improve this, either Schottky barriers or p^+-i-n^+ structures are generally used in the photovoltaic cells based on hydrogenated amorphous silicon. Photovoltaic cells using Schottky barrier have a simpler structure (n^+-i-Me structure). The metal, forming Schottky barrier, should have a high work function of electrons (Pt, Rh or Pd) since the undoped silicon layer has weak *n*-type conductivity (see Sec. 4.1). High work function of electrons from the metal causes formation of positive space charge in the semiconductor layer.

However, higher efficiency of photovoltaic cells has been achieved by using p^+-i-n^+ structures (Fig. 7.3(a)). There are two junctions (p^+-i and n^+-i) in this structure. Concentration of charged centers in *i*-type layer is relatively low due to the absence of doping. Therefore, in the case of film with thickness of 0.4–0.8 μm space charge region extends to the full thickness of *i*-type layer (Fig. 7.3(b)). Consequently, electric field exists in the whole *i*-type layer (Fig. 7.3(c)) and the charge carriers photo-generated in any part of *i*-layer are separated by an electric field that will ensure their efficient collection. Electrons and holes photo-generated in the n^+- and p^+-layers do not contribute significantly to the photocurrent of the cell because of low charge carriers' lifetime in heavily doped a–Si:H layers. Therefore, thickness of the n^+- and p^+-layers typically ranges from 8 to 20 nm.

The next question, most important from the standpoint of efficiency, is from which side (through the p^+- or n^+-layer) to bring sunlight to the cell's active region. Since the absorption of photons is more probable nearer to the illuminated surface, then in the case of illumination from the p^+-layer (as shown in Fig. 7.3(c) the path for electron transition in *i*-layer is longer than the path for hole transition. This is direct construction of photoelectric cell. If the cell is illuminated from the n^+-layer side (reversed construction of photovoltaic cell), holes pass along the longer path in *i*-layer.

The diffusion length of electrons is longer than the diffusion length of holes in hydrogenated amorphous silicon. From this point of view, the direct design of photovoltaic cells is preferable because it ensures better collection of photo-generated carriers at the contacts. However, it is necessary to consider one more

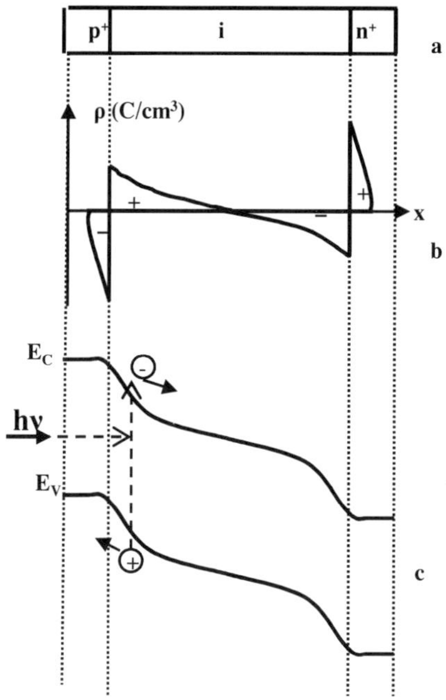

Figure 7.3. Schematic illustration of p$^+$-i-n^+ structure (a), distribution of electrical charge (b) and energy band diagram (c).

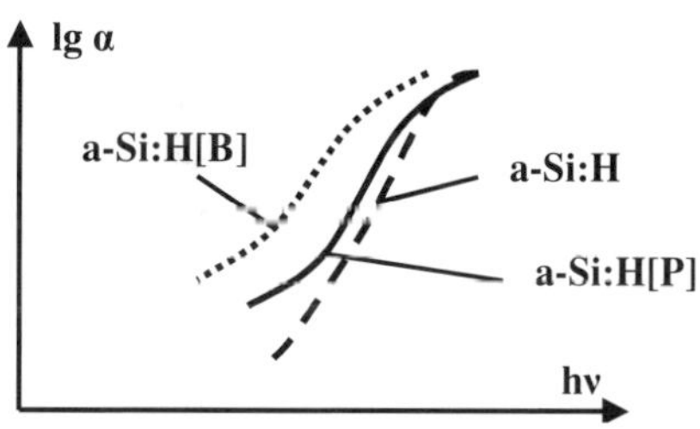

Figure 7.4. Schematic illustration of doping influence on spectral dependence of a–Si:H optical absorption coefficient.

factor when choosing the design of photovoltaic cells. Any doping increases optical absorption coefficient of a–Si:H. Moreover doping with boron increases the optical absorption coefficient of a–Si:H to a much larger degree than doping with phosphorus (Fig. 7.4). Thus, losses caused by photon absorption in a heavily doped layer in the case of direct design (illumination through boron doped layer) are higher in comparison with reversed design of the cell. Based on the above factors, designers of a–Si:H photovoltaic cells use both direct and reversed designs of the cells (Carlson, 1984). Examples of these designs are shown in Fig. 7.5.

In the case of a cell with direct design, it is possible to decrease the absorption losses in heavily doped p^+-layer by replacing p$^+$a–Si:H with p$^+$a–SiC:H. The latter has a larger band gap and is transparent to photons with energy equal to the width

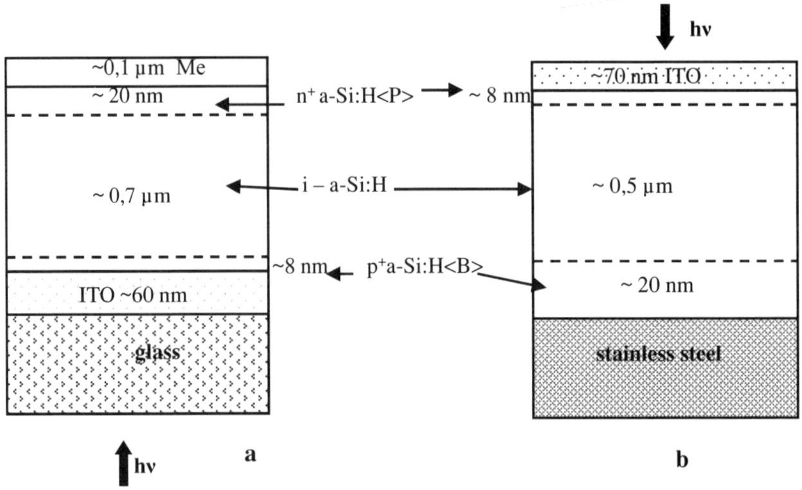

Figure 7.5. Direct (a) and reversed (b) constructions of the a – Si:H photovoltaic cells (ITO is indium tin oxide).

Table 7.1 Properties of hydrogenated amorphous silicon device quality films.

Property	i-a–Si:H	p-a–$Si_{1-x}C_x$:H[B]	n-a–Si:H[P]
Gap value, eV	1.8	2.0	1.8
Refractive index	4.3		
Absorption coefficient at λ= 600 nm, cm^{-1}	$4{\cdot}10^4$	10^4	$4{\cdot}10^4$
Conductivity, $Ohm^{-1}cm^{-1}$	10^{-10}	10^{-5}	10^{-3}
Photoconductivity, $Ohm^{-1}cm^{-1}$	10^{-5}		
Conductivity activation energy, eV	0.8	0.5	0.3
Hydrogen content, at percent	8 – 12	8 – 12	8 - 12
Density of defect states $eV^{-1}cm^{-3}$	10^{15}		

of a–Si:H gap. Properties of hydrogenated amorphous silicon device quality films are given in the Table 7.1 (Sherchenkov and Budagyn, 2007).

The above designs ensure efficient collection of photo-generated carriers on the contact electrodes. Let us consider implementation of other conditions for efficient conversion of solar energy into electricity in a photovoltaic cell.

The condition of absorption of significant part of solar radiation in the active region of the cell. As mentioned above, thickness of i-layer in a photovoltaic cell (0.4–0.8 μm) is determined by the requirement of the electric field existence throughout the volume of the layer. However, i-layer of this thickness does not ensure absorption of a considerable part of solar radiation, which has a broad spectral range from 0.3 to 1.0 μm. Only radiation with photon energy approximately equal to the semiconductor band gap is converted effectively into electricity (Fig. 7.6(a)). Photons with lower energy are not absorbed by semiconductor material. Photons with higher energy, as a rule, are absorbed in the heavily doped layer. Even in the case of high energy photon absorption in the active *i*-layer, a significant portion of its energy is spent on thermalization of generated electron and hole and is converted into heat (Fig. 7.2(a)).

The solution of this problem was implemented on the basis of cascade or multiple-junction solar cells. In the case of two a–Si:H cells, formed one on top of another as a sandwich, instead of one cell, the radiation that is not absorbed in the first cell may be absorbed in the second. This increases the efficiency of sunlight absorption (Fig. 7.6(b)). However, efficiency increases still occur only in a narrow region of the spectrum at photon energies close to the value of a–Si:H band gap. It is necessary to use semiconductors with various value of the band gap for the expansion of the spectral characteristics of the cell's optical absorption. In the case of a double-junction solar cell the upper element should have a wider, and the lower element should have a narrower band gap (Fig. 7.6(c)). In the case of triple-junction solar cell band gap of elements should decrease from the illuminated surface to the opaque electrode (Fig. 7.6(d)). This ensures efficiency of photovoltaic cell in a wider part of the solar radiation spectrum.

Solid solutions of silicon with other elements of the periodic table's fourth group (germanium, carbon, nitrogen) and microcrystalline silicon are used to create various band gap materials based on hydrogenated amorphous silicon, (see Secs. 5.2 and 5.3). This makes it possible to obtain semiconductor layers with band gap value from 1 to 3 eV. The structures of single and multiple-junction photovoltaic cells are listed in the Table 7.2.

The efficiency dependencies of solar cells on band gap values of the middle and bottom elements are shown in Fig. 7.7. The band gap of the top element is equal to 1.95 eV (Yukimoto, 1983). The figure shows that the maximum efficiency is achieved with the following gap values: $E_{g1} = 1.95$ eV, $E_{g2} = 1.45$ eV and $E_{g3} = 1.0$ eV. It should be noted that a necessary condition for maximum efficiency of multiple-junction photovoltaic cell is the equality of short-circuit currents of all its elements (assuming the equality of their Fill Factors). Since the band gap and resistivity of *i*-layers are different, layer thicknesses should be equal $d_{i1} = 0.2$ μm, $d_{i2} = 0.5$ μm and $d_{i3} = 1.0$ μm to fulfill this condition.

Another way to reduce optical losses in photovoltaic cells is to use reflective layers or the so-called light trapping. The idea of the method is that the special textured layers at the front and back surfaces of the cell (zinc oxide and silver layers in Fig. 7.6 for example) ensure multiple internal reflections. This leads to the increase of the optical path up to 50 times and a sharp growth in light absorption.

The condition of small value of internal photovoltaic cell resistance. As shown above, highly doped n^+ and p^+ layers of a–Si:H or μc–Si:H form contacts with electrodes in the photovoltaic cells. Therefore, the internal resistance of the cell

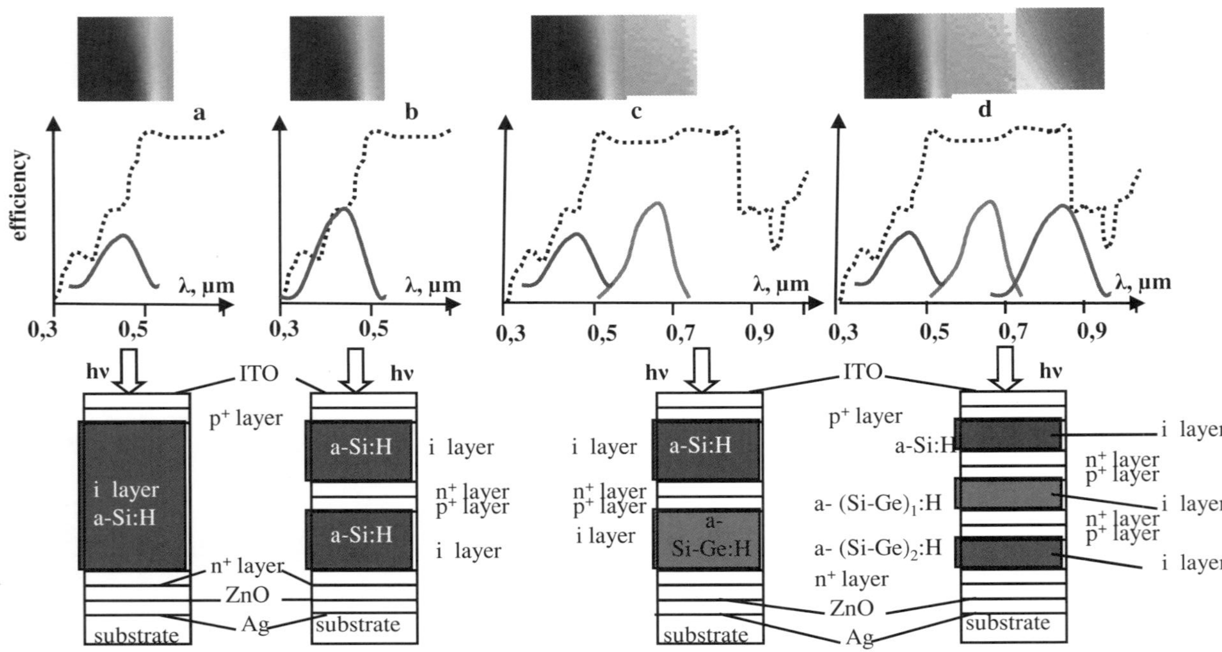

Figure 7.6. Schematic illustration of the spectral dependence of the efficiency(top) and structure(bottom) of multiple-junction solar cells: single-junction (a) double-junction with the same gap (b) double-junction with dual gap (c) and trible junction (d) (the dashed line is the spectrum of sunlight on the Earth's surface). For color reference, go to page 196.

Table 7.2 The structures of single and multiple-junction photovoltaic cells.

Single-	Double-junction		Triple-junction		
junction	Top	Bottom	Top	Middle	Bottom
a–Si:H	a–Si:H	a–Si:H	a–Si:H	a–Si:H	μc–Si:H
–	a–Si:H	a–Si_xGe_{1-x}:H	a–Si:H	a–Si_xGe_{1-x}:H	μc–Si:H
–	a–Si:H	μc–Si:H	a–Si:H	a–Si_xGe_{1-x}:H	a–Si_yGe_{1-y}:H

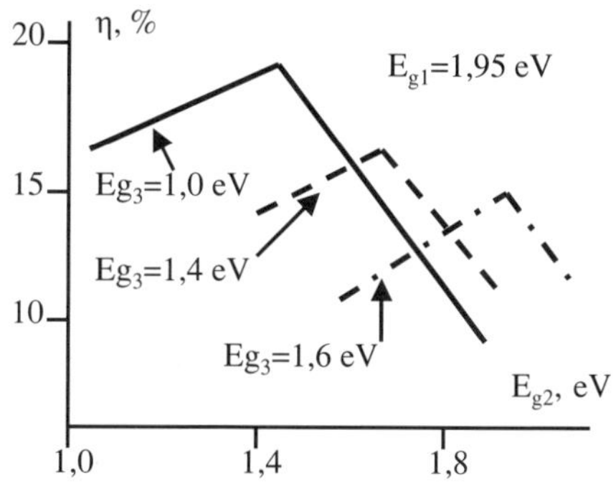

Figure 7.7. The solar cell efficiency dependencies on band gap values of the middle and bottom elements. For color reference, go to page 197.

should be small. This is confirmed in the case of small area cells. However, increasing photovoltaic cell's area leads to a sharp rise in internal resistance and a corresponding decrease in short-circuit current and efficiency. The reason for this effect, called the size effect, becomes clear from the results of a simple experiment (Kuwano and Ohnishi, 1983). a–Si:H photovoltaic cell 1 × 1 cm in size was made on a glass substrate of 10 × 10 cm size coated with a transparent indium-tin oxide (ITO) electrode. Metal contacts to the transparent electrode were formed at the corners of the substrate (Fig. 7.8(a)). Length of current path (l) through the

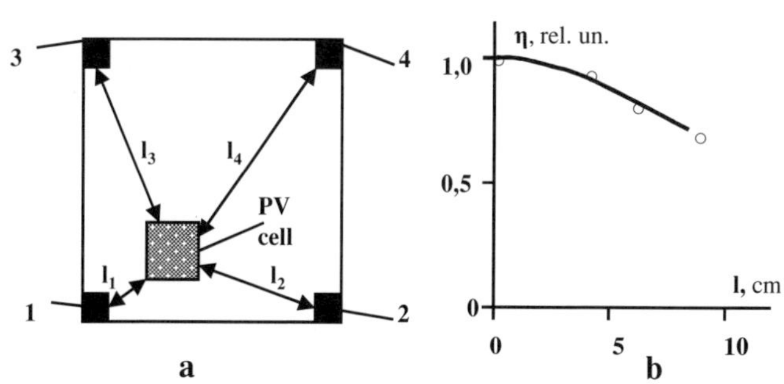

Figure 7.8. Illustration of the photovoltaic cell size effect (see text).

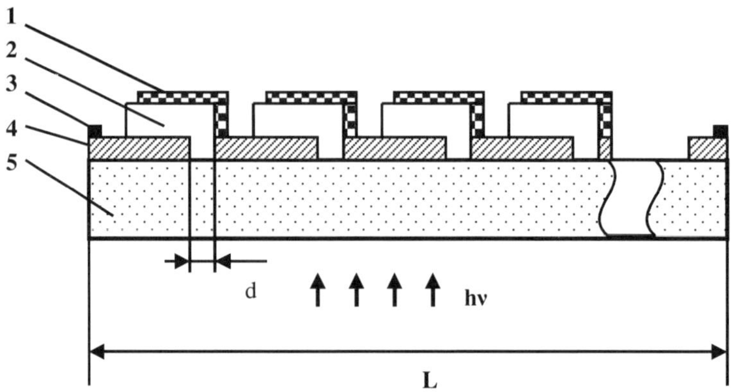

Figure 7.9. Design of integrated photovoltaic module (1 is upper metal electrode, 2 is photovoltaic cell, 3 is metal contact, 4 is transparent ITO electrode and 5 is glassy substrate).

transparent ITO electrode was varied by attaching the cell through various contacts (1–4). The resistance of ITO electrode varies from 10 to 30 Ohm/□. Therefore, increasing the length of the current path causes a significant increase in internal resistance of the cell and reduces its efficiency (Fig. 7.8(b)).

The influence of size effect on photoelectric characteristics can be reduced by reducing the length of the current path through the transparent electrode. To achieve this, the integral design of photovoltaic modules was developed (Fig. 7.9). Production process for these modules is as follows. Rectangular transparent ITO electrodes 4 are deposited on the substrate 5. Spacing between the electrodes is equal to d. Photovoltaic cells 2 are formed after that. The final step is the formation of metal contacts to the ITO electrodes (3) and the cell upper electrodes (1). Moreover, the upper electrode of the first cell is connected with the ITO electrode of the second cell along all cell length. The upper electrode of the second cell is connected with ITO electrode of the third cell and so on. Thus all the cells in photovoltaic module are connected in series. As one can see, the length of current path through transparent electrodes is sharply decreased in this case and hence power losses are reduced as well. Replacement of one cell to fifteen cells formed in the same area reduces the power losses in the transparent electrode by approximately 100 times. However the active module area decreases with increasing number of cells due to the rise of total cell interconnection area. Therefore, optimization of the number of cells in the module is required.

Efficiency of integrated module is given by

$$\eta_M = \eta_0\{1 - [(N+1)d/L]\}[1 - (P_{los}/P_\Sigma)] \tag{7.3}$$

where η_0 is efficiency of small area cell in which the power losses in the transparent electrode are neglected, N is number of cells in the module, d is spacing between the electrodes, L is module width, P_{los} is power losses in the transparent electrode and P_Σ is total module power.

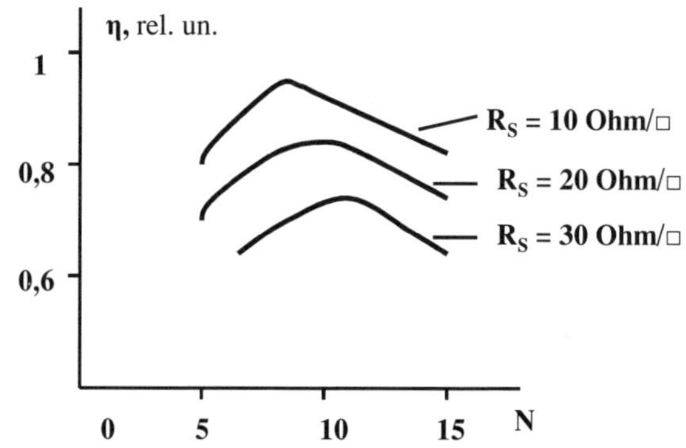

Figure 7.10. Dependence of module efficiency on number of cells (R is surface resistance of transparent electrode).

The results of module efficiency calculation in the case of different values of the transparent electrode surface resistance are shown in Fig. 7.10. Module size was 10×10 cm and spacing between electrodes $d = 1$ mm. As seen from the figure, the optimal number of elements in the module increases with increasing resistance of the transparent electrode.

Integral design of photovoltaic module ensures the implementation of condition for high value of open circuit voltage as well. Open circuit voltages of all cells are summed up due to serial connection of cells in the integral module. As a result, the output voltage varies from 70 to 140 V at 1 m module width.

Guaranteed service life of photovoltaic modules based on hydrogenated amorphous silicon is assumed to be 20 years. The factors limiting the service life can be divided into two groups. The first group includes the design and technological factors, such as the reliability of element hermetical sealing, the choice of electrode material to minimize its diffusion in the semiconductor and others. The influence of these factors can be reduced by improving the design and manufacturing process. The second group includes factors that depend on fundamental properties of semiconductor material.

The Staebler–Wronski effect in hydrogenated amorphous silicon belongs to the latter (Staebler and Wronski, 1977). The effect is to reduce both the dark- and photo-conductivity of hydrogenated amorphous silicon under long-term effect of illumination. It was found that the conductivity changes are due to the formation of new metastable localized states in the gap of the semiconductor under the influence of illumination. The dependence of the localized states density in the gap of a–Si:H on the illumination time (Sherchenkov and Budagyn, 2007) is shown in Fig. 7.11. The nature of these metastable states is not completely clear. Usually their appearance is associated with the formation of dangling bonds on silicon atoms due to recombination of photo-generated carriers and the interaction of these dangling bonds with hydrogen atoms. The effect is reversible in nature. Annealing at 150–200°C for an hour leads to a complete renewal of the dark- and photo-conductivity.

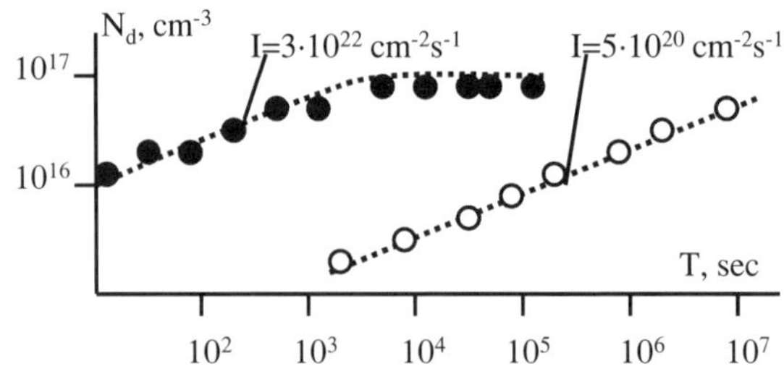

Figure 7.11. The dependence of the localized states density in the gap of a–Si:H on the illumination time at various intensities of illumination.

Figure 7.12. Uni-Solar amorphous silicon photovoltaic power system in California, USA (4800 laminates 5.4×0.4 m^2, 500 kW, Gregg *et al.*, 2005). For color reference, go to page 197.

The Staebler–Wronski effect is observed in photovoltaic cells based on hydrogenated amorphous silicon as well. As a result, efficiency is reduced by 10–20% from its initial value during the first year of operation. After that the efficiency is stabilized and only slightly changes during the subsequent operation. At present stabilized efficiency of commercially produced photovoltaic modules based on hydrogenated amorphous silicon is 8–10%.

An example of large scale amorphous silicon ground mounted photovoltaic system is shown in Fig. 7.12 (Gregg *et al.*, 2005).

7.2 SWITCHING AND MEMORY DEVICES ON THE BASIS OF CHALCOGENIDE ALLOYS

7.2.1 Threshold Switching Effect in Chalcogenide Glassy Semiconductors

Chalcogenide glassy semiconductors demonstrate threshold switching effect or rapid transition from high resistivity state to high conductivity state and back. The

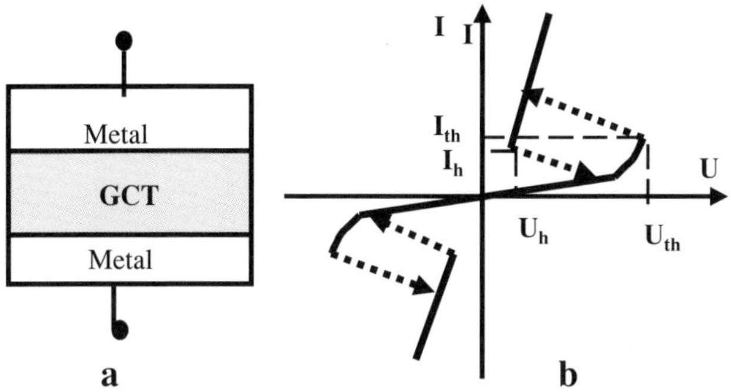

Figure 7.13. Structure (a) and voltage-current characteristic (b) of threshold switcher on the basis of chalcogenide glassy semiconductor.

first discovery of switching in chalcogenide glasses was independently made by S. Ovshinsky and B. Kolomiets in 1963. Despite numerous studies, a unified theory of the switching effect has not yet been created. There is a number of coexisting hypotheses and models to explain this effect.

A typical chalcogenide glass threshold switching device is about 1 μm thick film sandwiched between two metal electrodes (Fig. 7.13(a)). The device has a symmetrical voltage-current characteristic with negative differential resistance (Fig. 7.13(b)).

The initial state of device is a high-resistance. At low applied voltages the conductivity varies linearly, and the resistance is about 10^5 Ohms. When the electric field strength reaches about 10^4 V/cm linear law is replaced by non-linear (exponential) low, which corresponds to the space charge limited current. Switching the device in low resistivity state occurs when a voltage rises to the threshold value (U_{th}). The threshold voltage is linearly dependent on film thickness (Fig. 7.14). It conforms to consistent voltage distribution across the film

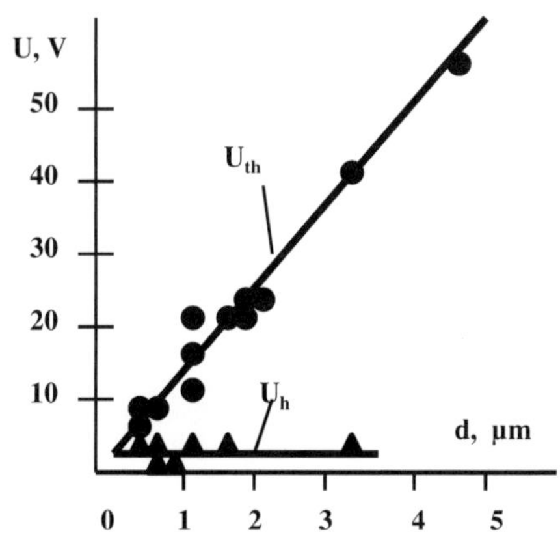

Figure 7.14. Dependences of threshold voltage (U_{th}) and holding voltage (U_h) on chalcogenide glassy film thickness.

and constant value of threshold electric field (approximately 10^5 V/cm). The device resistance decreases by 5–6 orders of magnitude. Dynamic resistance in low-resistance state is typically about 1 Ohm. Reducing the current in the low-resistance state to a critical value (I_h) causes the device to switch to its original high-resistance state.

The current is proportional to the area of electrodes in the high-resistance state, but in the low-resistance state it does not depend on the area of electrodes. This means that the current channel (filament) with small cross section and with high current density is formed in the low-resistance state. Since the current flows through the small cross section channel in the low-resistance state, the maximum current of the switch, as a rule, does not exceed several mA.

Holding voltage of low-resistance state (U_h) does not depend on the thickness of semiconductor film (Fig. 7.14). This indicates that the most part of voltage applied to the device drops at the contacts, but not in the bulk of the semiconductor. As one can see from Fig. 7.14 there is a critical thickness of the semiconductor film (approximately 0.2 μm) at which the threshold voltage and the holding voltage are equal to each other. Consequently, when film thickness is smaller than this value switching is impossible in this type of devices. It should be added that the threshold voltage decreases with increasing temperature and with increasing pressure, but does not depend on the illumination. Under illumination only the threshold current increases at a constant value of the threshold voltage.

An important characteristic of switches is their switching speed. The time response of threshold switch to a voltage pulse is shown in Fig. 7.15. The time of switching from high-resistance in low-resistance state (switch-on time) consists of two components: delay time (t_d) and time of the switching process (t_s). Delay time is the time that elapses from the applying of a voltage pulse equal or greater to threshold value and before a sharp increase in current through the device. The delay time depends strongly on the applied voltage, varying from 10^{-6}–10^{-7} sec with $U = U_{\text{th}}$ to about 10^{-9} sec with voltage increasing. Time of the switching process does not exceed 10^{-9}–10^{-10} sec.

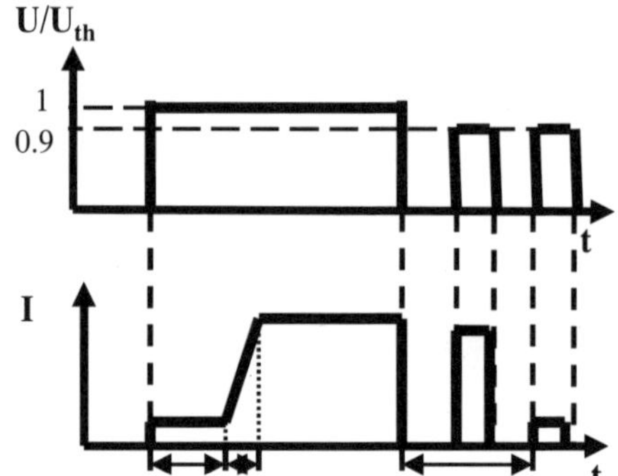

Figure 7.15. The time response of threshold switch to a voltage pulse (see text).

The renewal time of high-resistance state (switch-off time t_{sb}) is defined as the minimum time that has elapsed since the end of the pulse that switched the device into the low-resistance state, after which a pulse with amplitude of 10% less than the threshold voltage does not switch the device into low-resistance state. This time is usually about 10^{-6} seconds and increases with the thickness of chalcogenide glassy film.

The mechanism of threshold switching in chalcogenide glassy films is not fully understood. Existing hypotheses can be divided into three groups:

- thermal hypotheses, based on the consideration of thermal processes in the material,
- hypotheses based on the consideration of electronic processes,
- mixed electron-thermal hypotheses.

Historically, thermal hypotheses based on the theory of thermal breakdown were developed first. According to these hypotheses, the current flowing through the sample raises the temperature of the material, which leads to an increase in electrical conductivity and current value and to the heating of the sample. Insufficient heat removal causes an avalanche process that leads to loss of electrical strength of the sample. At first approximation, the process described above is consistent with the experimental dependencies, observed at threshold switching in the chalcogenide glassy semiconductors. Channel (filament) with high current density is formed in both cases. At the qualitative level theory of thermal breakdown explains the presence of switching delay time (time required for pre-heating of the sample), the presence of high-resistance renewal time (time required for cooling of the sample) and the dependence of delay time on the voltage (higher voltage causes higher current and more rapid heating). However, quantitative analysis showed that the thermal hypotheses cannot explain a lot of experimental dependencies.

Temperature dependencies of threshold voltage for thick (300 μm) and thin (2 μm) samples are shown in Fig. 7.16(a). In the case of thick sample, threshold

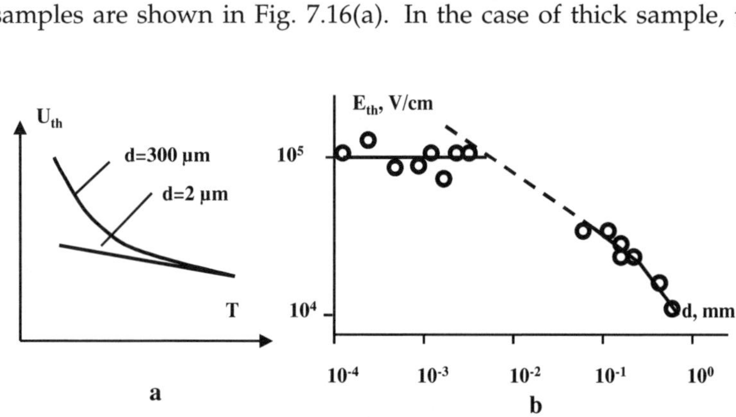

Figure 7.16. Schematic illustration of threshold voltage temperature dependences for samples with different thickness (a) and dependence of threshold electric field on sample thickness (b).

voltage decreases sharply with increasing temperature. This is both qualitatively and quantitatively consistent with the thermal breakdown theory. However, in the case of a thin sample the above mentioned dependence is much weaker and cannot be explained from the standpoint of the thermal hypothesis.

The dependence of the electric field at which switching occurs (threshold field) on the thickness of the sample is shown in Fig. 7.15(b) (Kolomiets *et al.*, 1969). The threshold field decreases with increasing thickness in the case of samples with thickness of 100 μm and more. In terms of thermal breakdown this fact is well explained by heat transfer deterioration. However, threshold field does not depend on the thickness of the sample of less than 10 μm thick. Moreover it was found that threshold field in thin samples depends on the electrode material. Besides that, the influence of the electrode material is not due to its thermal conductivity, but due to the value of metal-semiconductor work function. All these facts do not fit into the thermal switching hypothesis.

Thus, quantitative analysis allows to conclude that the model of reversible thermal breakdown is valid only for samples with semiconductor layer thickness of hundreds microns or more. Since film thickness of less than one micron is used in devices, it is necessary to consider other models of threshold switching.

Electronic hypotheses use various effects and processes to explain threshold switching. These are double injection of carriers from the electrodes, electronic phase transitions, tunneling effect, avalanche multiplication, switching in many *p-n* junction structures and others (Kostylev and Shkut, 1978). Consider as an example, the hypothesis of double carrier injection.

When voltage is applied across the device, initially the electric field is distributed uniformly in the semiconductor film. Consequently, the slope of the semiconductor energy bands is constant at this moment (Fig. 7.17(a)). The injection of electrons and holes from electrodes into semiconductor occurs under the electric field. However, due to the high density of traps (localized states) in the band gap of semiconductor, injected carriers are trapped. Filling the traps, the charge carriers are moving deeper into the semiconductor film and reach its central part. In the central part of the film both electron and hole traps are filled and compensation of electrical charges takes place. This leads to a redistribution of the electric field in the film (Fig. 7.17(b)). Over time, the central region of the film with high mobility of charge carriers increases and fills almost the whole of the volume between the electrodes. The electric field is concentrated in narrow near-electrode regions (Fig. 7.17(c)), which leads to tunneling of carriers from the electrodes into the semiconductor and the device switches to low-resistance state.

The model explains presence of delay time, weak dependence of the voltage across the device in the low-resistance state on the film's thickness and a number of other experimental results. However, some quantitative estimates on the basis of this model do not correspond to the experimental results (quantitative estimate of the delay time in thin films, for example).

Mixed electron-thermal hypotheses are considering the contribution of both electronic and thermal processes to the threshold switching. In general terms,

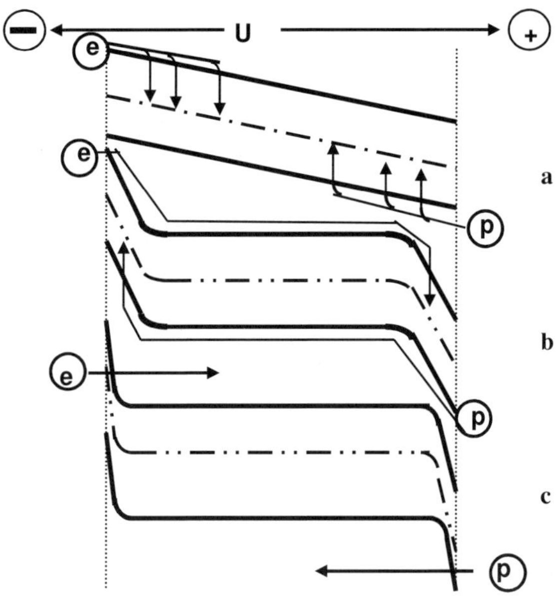

Figure 7.17. Switching process after double injection model (see text).

their idea is the following. Free charge carriers obtain additional energy from the electric field. Part of this energy is transferred to the atoms due to thermal scattering of charge carriers, which leads to an increase in temperature of the structural network. Temperature increase leads to a rise in concentration of free carriers. Avalanche growth of this process switches the device into low-resistance state. In other words, electronic hypothesis consider the electrical conductivity dependences on the electric field and on the temperature independently of each other ($\sigma = f[F, T]$). But in the case of electron-thermal hypotheses temperature depends on the electric field as well ($\sigma = f[F, T(F)]$).

Let us consider the percolation model of threshold switching. It is known that at the same width of the gap and, consequently, for the same carrier concentration, amorphous semiconductors' electrical conductivity is many orders of magnitude less than the electrical conductivity of the crystal. The reason for this is that most of the time charge carriers are in localized states and do not take part in conduction. At equilibrium conditions, generation of charge carriers from localized states and recombination of free charge carriers have the same rate. The electric field increases the rate of carrier generation and concentration of free carriers. Regions with higher concentration of free carriers appear as a result. Electric field is redistributed in the space between electrodes. Its value increases between the regions with higher carrier concentration. It stimulates local emission of charge carriers and leads to formation of percolation current filament between electrodes. This percolation filament may be destroyed or transformed into a stable current channel. In the latter case, due to additional heat effect, the area of channel increases and switching into low-resistance state takes place.

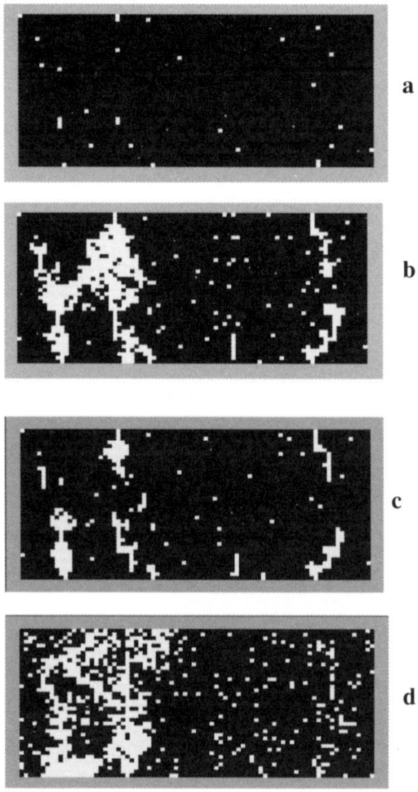

Figure 7.18. Simulation of current filament formation (light spots indicate the regions with higher carrier concentration): (a) corresponds to the moment of voltage application, (b), (c) and (d) are the model after 60, 71 and 200 iterations.

The simulation results of threshold switching in the frame of percolation model are shown in Fig. 7.18 (Voronkov *et al.*, 2006). Light spots in the figure indicate the regions with higher carrier concentration. After 60 iterations percolation current filament is formed between electrodes (Fig. 7.18(b)). However, the filament is destroyed after 71 iterations (Fig. 7.18(c)) and stable current channel is formed due to increase of channel area only after 200 iterations (Fig. 7.18(d)).

7.2.2 Bistable (memory) Switching in Chalcogenide Glassy Semiconductors

Memory switching effect in chalcogenide glassy semiconductors was discovered simultaneously with the effect of the threshold switching. The structure of memory switch is completely the same as the structure of the threshold switch (Fig. 7.13(a)).

Initially, the current-voltage characteristic of the memory switch repeats current-voltage characteristic of the threshold switch. The initial resistance of the device is high. The device switching into low-resistance state occurs under thresh-

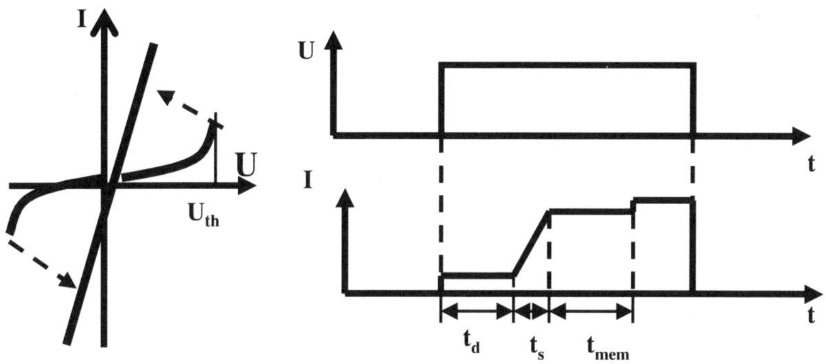

Figure 7.19. The current-voltage characteristic (a) and the time response to voltage pulse (b) of the memory switcher.

old voltage action. Reducing the current through the device below the holding current value immediately after switching leads to its return into a high-resistance state, just as in the case of the threshold switcher. However, the low-resistance state is retained even when the external voltage is turned off, if the device was held in the low-resistance state for about 10^{-3} sec (Fig. 7.19(a)). Once locked in the low-resistance state the device can be returned to the original high-resistance state by applying a short intense "erase" current pulse of either polarity.

The time response of memory switches to voltage pulse is shown in Fig. 7.19(b) In comparison with the threshold switches, there is one additional parameter in this case. It is time of memorizing low-resistance state (lock-on time, t_{mem}). This time is about 10^{-3} sec. Lock-on low-resistance state is accompanied by a slight increase of the current through the device (Fig. 7.19(b)). Thus this type of switch allows storing information without energy expenditure.

The difference between the threshold and bistable switch characteristics is determined by different chemical composition of chalcogenide glassy semiconductors. Materials with high values of crystallization activation energy are used in threshold switches. Due to this, the semiconductor film remains in the glassy state during device operation. Composition of these materials is in the middle of the glass formation regions of the chemical systems (see Fig. 1.8). In contrast, memory switches are prepared on the basis of chalcogenide glasses with low activation energy of crystallization. The composition of these materials is near the borders of glass formation regions.

Low-resistance state in memory switches is realized due to phase separation and crystallization of glassy material in current channel under the influence of Joule heating. Crystallization of current channel between the electrodes leads to lock-on of low-resistance state, since resistivity of the crystalline phase is less than the resistivity of glass by the orders of magnitude. Two different mechanisms of material crystallization process in the current channel are possible depending on parameters of current pulses and glass composition: glass — crystal transition and

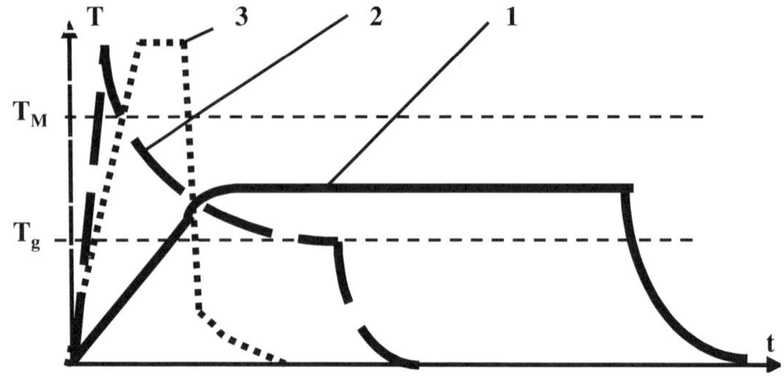

Figure 7.20. Dependencies of current channel temperature on time of low-resistance state recording in the case of glass-crystal transition (curve 1), glass-melt-crystal transition (curve 2) and on time of high-resistance state recording (curve 3).

glass — melt — crystal transition. In the first case the temperature of material in the current channel increases under current pulse action to a value above glass transition temperature but below melting temperature of the crystalline phase (curve 1 in Fig. 7.20). The result is crystallization of the material in the current channel between electrodes.

In the second case temperature of the channel increases to a value higher than melting temperature of the crystalline phase and material turns into liquid state. The pulse quench must be sufficiently slow to ensure material crystallization during cooling (curve 2 in Fig. 7.20). The comparison of two versions shows that the glass-crystal transition is easier from kinetic point of view and, consequently, more reproducible. Lock-on time is equal to about 10^{-3}sec in this case. The advantage of the glass-melt-crystal transition is lower lock-on time of low-resistance state.

Back switching into high-resistance state can be achieved only through the transition crystal-melt-glass. Implementation of this transition is carried out by a short current pulse with sharp quenching. The current value should provide increase in channel temperature above the melting point of crystalline phase. The sharp quenching of the pulse leads to rapid cooling of the channel and vitrification of the melt (curve 3 in Fig. 7.20) and the device switches into high-resistance state.

The choice of material is a key issue in designing any device. As noted above, materials with a sufficiently rigid and stable structural network (with high crystallization activation energy) are required for threshold switches. In contrast, materials with a fairly easily changeable atomic structure are used in bistable switches. The first industrially produced threshold switches were created on the basis of $Si_{12}Ge_{10}As_{30}Te_{48}$ (the criterion of structural modification efficiency is equal to 0.95, see Sec. 4.4.2) and bistable switchers were created on the basis of $Ge_{15}Te_{81}As_4$ (the criterion of structural modification efficiency is equal to 1.68). However, these requirements are necessary but not sufficient, since all switch parameters depend on the chemical composition of chalcogenide glassy semiconductor. For example, threshold electric field varies from 10^4 to 10^6 V/cm depending on the chemical composition and on the band gap value of the material. In summary, the following requirements to materials for memory switches may be formulated:

- the chemical compositions of the material has to be near the boundary of glass formation region,
- the material has to have minimum value of difference between temperature of crystallization and glass transition temperature,
- the material has to have maximum value of difference between temperature of melting and temperature of crystallization,
- the band gap of the material should not exceed 1 eV.

Along with absolute values of switching characteristics the stability of the characteristics during device operation is very essential. There are two types of switch characteristic changes: consistent change of the parameters (threshold voltage, in the first place) during the first cycles of switching (called the forming process) and statistical variation of parameters during device operation (Fig. 7.21).

Presence of the forming process is explained by the strongly non-equilibrium process of film formation. Atomic structure of as-prepared films corresponds to the non-stable state, located on the slope of the meta-stable state minimum on the configuration diagram (point N_2 in Fig. 1.4). Therefore, threshold voltage of as-prepared devices is generally about 30% larger than threshold voltage of the formed devices. During the first 10–20 switching cycles the system moves to meta-stable state due to structural relaxation. After this threshold voltage is stable within the limits of statistical variations. In order to eliminate the need for forming process, it is necessary to obtain films with the structure corresponding to meta-stable state. Some technological methods, consisting either in vapor-liquid-glass mechanism of film growth (Kostylev and Shkut, 1978), or in special heat treatment after device fabrication have been developed to solve this problem.

In contrast to the forming process, the statistical variation of electrical parameters can not be avoided and is determined by the material's atomic structure. It has been shown in section 4.4.5 that statistical variations of threshold voltage depend

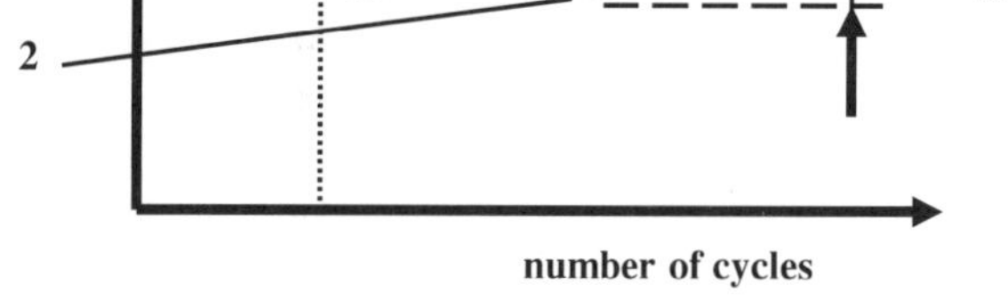

Figure 7.21. Schematic illustration of threshold voltage changing during device operation (1 is range of forming process, 2 is range of statistical variation).

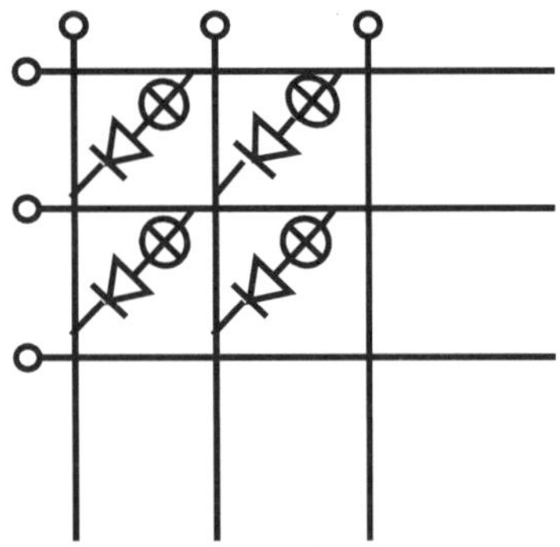

Figure 7.22. Diagram of non volatile memory integrated circuit on the basis of chalcogenide glasses.

on the value of criterion of structural modification efficiency (see Fig. 4.12). This dependence can be written as follows

$$\Delta U_{th} = \alpha(CSME)^{\beta} + \Delta U_0 \tag{7.4}$$

α and β are coefficients, *CSME* is criterion of structural modification efficiency and ΔU_0 =1–2%. Thus, it is necessary to choose chalcogenide glasses with smaller value of this criterion in order to minimize statistical variation.

Integrated circuits of nonvolatile memory based on memory switches were developed and manufactured in the U.S. and the USSR in the 1970s. Diagram of these circuits is shown in Fig. 7.22. There is a matrix of mutually perpendicular buses, each cell of which contains a serially connected memory switch described above and a silicon diode.

However, as it was soon discovered, these integrated circuits were not sufficiently reliable. In the process of information storage and reading some of the cells switched from one state to another spontaneously. The reasons for such behavior were related to both operational conditions and material properties. After switching from low-resistance to high-resistance state, the crystalline channel in the device did not disappear completely. Crystalline inclusions remained in the glassy matrix. They had small influence on the device resistance, but could play the role of spontaneous crystallization centers in the process of operation. As a result, spontaneous switching to low-resistance state occurred. In the case of the reverse situation, crystalline channel connecting the electrodes in a low-resistance state could destroy in a certain "weak" spot during the operation process. As a result spontaneous switching to high-resistance state occurred. For this reason, non volatile memory integrated circuits based on chalcogenide glassy semiconductors were not developed further and their production was discontinued. But thirty years later, these principles of information recording have found application in new devices, which will be discussed in the next section.

7.2.3 The Second Generation of Phase-change Memory Devices

In the early 2000s new memory devices based on materials of Ge–Sb–Te (GST) system were developed by the Ovonyx Company. This memory was first called Ovonic Unified Memory (OUN), and later Phase Change Memory (PCM). List of companies involved in the development of PCM-memory integrated circuit includes Intel and IBM (USA), BAE Systems and STMicroelectronics (Europe), Samsung (Korea), Elpida Memory (Japan) and others (Weiss, 2005). The developers believe that phase change memory could replace the traditional flash memory, static random access memory (SRAM) and dynamic random access memory (DRAM).

The advantages of this type of memory are numerous. There is a possibility to significantly increase the density of information recording. Ability to rewrite data without erasure provides recording acceleration approximately by factor of 30 compared with flash memory. Phase change memory production has fewer processing steps than flash memory.

2 Mb and 4 Mb radiation hard phase change random access memory (RAM) integrated circuits were created by BAE Systems. Samsung Electronics Company has demonstrated 64 Mb phase change random access memory integrated circuit based on 0.18 μm technology with the following parameters: cell size is $0.56\,\mu m \times 0.9\,\mu m = 0.504\,\mu m^2$, chip size is $6.34\,mm \times 11.04\,mm = 70\,mm^2$, supply voltage is 3.0 V, reset current is 0.8 mA, operation time is 60–100 ns, endurance is 10^9 cycles and retention time is 2 years at 85 °C (Yeung et al, 2005). And since June 2009 the company began mass production of 512 MB phase change RAM chips. The next step is production of 1GB chips.

Let us compare the new memory elements and memory elements based on chalcogenide glassy semiconductors, described in Sec. 7.2.2. The principle of recording in both cases is the same: reversible phase transitions glass ↔ crystal and associated changes in electrical conductivity of material. Therefore new memory elements can be called the second generation of phase-change memory. There are no significant differences in the matrix architecture as well. Decoupling diode (see Fig. 7.22) is replaced by a thin film transistor. The latter allows realization of end-to-end thin-film technology in manufacture of memory integrated circuit. The structure of the switch has not changed either. It is a sandwich: conductive electrode — chalcogenide alloy — conductive electrode (Fig. 7.23).

Along with the common features described above, new memory elements have a number of significant differences. The first difference lies in the fact that the chalcogenide alloy is in the crystalline state, and not in the glassy state. Hence initially the device has the low-resistance state. The second difference lies in the structural dimensions of the memory element. As seen from the Fig. 7.23(a), the dimensions of element parts are in the range of tens or hundreds of nanometers, which is an order of magnitude smaller than 1st generation devices. Smaller size comes out of application requirements (high density) but has an impact on the physics of device operation. The next difference is in the design of the bottom electrode.

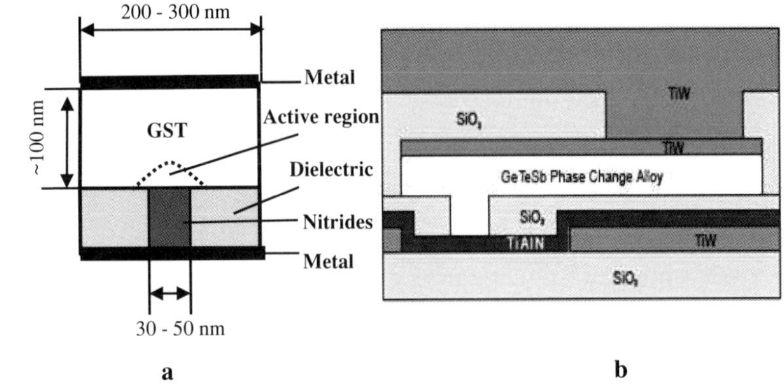

Figure 7.23. Phase change memory switch structure (a) and example of design (b).

The top and bottom electrodes in the 1st generation of memory elements had, as a rule, the same area. In contrast, in the case of the 2nd generation of memory elements the bottom contact to the chalcogenide alloy has an area several times smaller than the top contact (see Fig. 7.23(a)). Yet another difference lies in the chemical composition of chalcogenide alloy. The same materials of Ge–Sb–Te and Ag–In–Sb–Te systems are used for both the electrical memory and for the optical disc storage (see Sec. 6.2.3). These materials ensure stability of non-crystalline phase of up to 10 years on the one hand, and crystallization under the recording pulse action for 1–10 ns on the other hand.

As mentioned above, initially the device is in a low-resistance state. Switching into high-resistance state is carried out by applying a current pulse with a sharp rear edge. Phase change takes place due to Joule heating and following vitrification of the chalcogenide alloy. The highest heating of the material occurs in the area adjacent to the back electrode (active region in Fig. 7.23(a)), because the current density is greatest in this area. As a result, semispherical volume of non-crystalline phase is formed over the back electrode and the resistance of the device increases sharply. Radius of the non-crystalline region as a rule (defined by application requirements) does not exceed 50 nm. In the reverse switching longer but smaller in absolute value current pulse is applied to the device. Under current pulse processes of threshold switching, current channel formation and the subsequent crystallization of glassy drop are taking place. The device switches to low resistance state (Popov, Savinov and Voronkov, 2006). Thus, instead of crystallization and melting of filaments between the electrodes, which occurred in the first generation of phase-change memory devices (Sec. 7.2.2), in these devices the phase transitions occur in the bulk active region (Fig. 7.23(a)), which significantly increases their reliability. The dependence of device resistance on recording current is shown in Fig. 7.24 (Lowrey *et al.*, 2004).

Typically, the mechanism of switching from low-resistance to high-resistance state in the second generation of phase-change memory devices is explained on the basis of phase transitions crystal – melt – glass, similar to the first generation

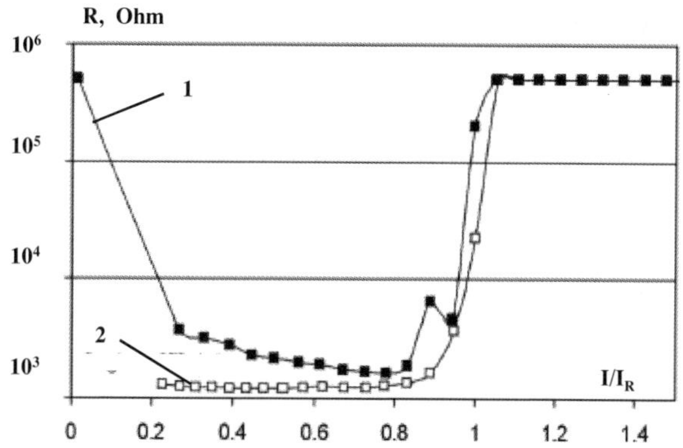

Figure 7.24. Dependence of device resistance on recording current (I_R is recording current of high-resistance state) in the case of initial high-resistance state (curve 1) and initial low-resistance state (curve 2).

(curve 3 in the Fig. 7.20). However as noted in section 6.2.3, fast and stable phase-change transitions in Ge–Sb–Te alloys are difficult to describe quantitatively by reference to conventional melting, amorphization and crystallization processes. Kolobov *et al.*, 2004 concluded that phase changes in these materials occur due to "umbrella-flip" structural transformation (see Fig. 6.27) without material melting in a conventional sense. An indirect confirmation of this conclusion is the existence of the forming process in the low-resistance state of as-prepared switches. This means that the free energy of the material in the low-resistance state does not correspond to the minimum in the configuration diagram (see Sec. 1.2) (Popov, 2009).

As noted above, two groups of chalcogenide alloys are used in phase-change memory elements, namely the materials of Ge–Sb–Te system and the materials of Me–Sb–Te systems (where Me is metal). The first are pseudo-binary GeTe–Sb_2Te_3 compositions. These include the $Ge_2Sb_2Te_5$(GST 225 is the most widely used material) and some other materials as $Ge_1Sb_2Te_4$ (GST 124) and $Ge_1Sb_4Te_7$ (GST 147). The governing process of glass-crystal phase transition in all of these materials is nucleation, since the speed of this process considerably exceeds the rate of crystal growth. Materials close to GeTe in chemical composition possess high thermal stability (high crystallization temperature) but low rate of crystallization. On the contrary, materials close to Sb_2Te_3 in chemical composition have high crystallization rate but low thermal stability. Improved properties can be achieved by chemical modification of GST materials. For example, addition of a few atomic percent of nitrogen in the material increases crystallization temperature and resistivity of the crystalline phase. Addition of oxygen leads to an increase in resistance ratio of high- and low-resistance states of the memory element (Lacaita and Wouters, 2008).

The second group of chalcogenide alloys is metal-modified $Sb_{70}Te_{30}$ compound. Additions of indium $[In_x(Sb_{70}Te_{30})_{1-x}]$ and silver and indium

$[Ag_xIn_y(Sb_{70}Te_{30})_{1-x-y}]$ (called AIST) are generally used. In contrast with GST, glass-crystal phase transition in these materials is due to growth of crystalline phase from the boundaries of the active region, rather than nucleation.

The ability to reduce sizes of phase-change memory cells and increase the density of information recording (scalability) is a very important factor in terms of competition with existing Flash and other types of memory. The scaling of phase-change memory includes (Lacaita and Wouters, 2008):

- scaling of the thermal profiles in order to avoid thermal disturbances while shrinking cell separation,
- scaling of the program current and voltage,
- preservation of the material's characteristics with miniaturisation.

This analysis shows that scaling potential of phase change memory is significantly superior to other types of memory.

Moreover, integrated circuits of phase change memory are quite easily produced by thin film technology. This allows to place several integrated circuits one above the other (3D technology) and thus to increase the density of information recording by several times.

Another approach to solve the problem of density is to store more than two levels (zero and one) per cell. Phase change memory is ideally suited for this task. As mentioned above, switching from low-resistance to high-resistance state leads to formation of glassy phase in the active region of the cell. The thickness of the formed glassy region depends on the programming current. In turn, the resistance of the cells in high-resistance state is determined by the thickness of the glassy region. Consequently, varying the program current should lead to a change in cell resistance. The results of this experiment are shown in Fig. 7.25. As seen from the figure, instead of two states (high-resistance and low-resistance), more than a dozen states with different resistance have been recorded.

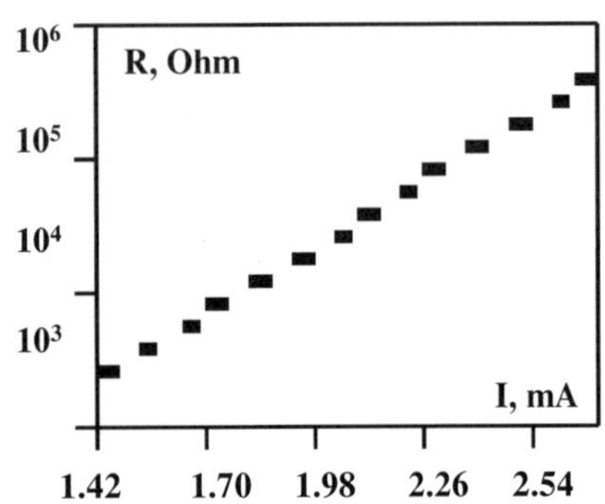

Figure 7.25. Cell resistance dependence on magnitude of programming current.

7.3 SILICON THIN FILM TRANSISTORS

Metal-oxide-semiconductor field effect transistors (MOSFETs) based on single-crystal semiconductors are the most common types of semiconductor devices at present. The question is: if this type of devices has been successfully produced on the basis of single crystals for a long time, why is it necessary to develop similar devices based on disordered semiconductors? As it turns out, along with the general advantages of devices based on disordered semiconductors (lack of fundamental size limitations, low cost, etc.) there are tasks that can not be fulfilled by using single-crystal devices. These tasks include creation of flat liquid crystal displays.

Creation of the screen on the basis of liquid crystals is not a big problem. A flat screen consists (Fig. 7.26(a)) of the two glass plates, one of which is covered with a transparent electrode (ITO) completely, and a similar electrode is formed as cells on the other plate. The gap between the plates is filled with liquid crystal. Optical properties of liquid crystal are changed by the electric field in the cells, where voltage is applied. The problem lies in the control of the screen. In the case of resolution of 10 dots per millimeter, the screen 10×10 cm should have 10^6 cells, each of which must be independently controlled. This means that there has to be a million connections between the control circuit driving screen cells and the screen itself. Reducing the number of connections is possible only in the case of placement of driving circuit directly on glass substrate of the liquid crystal screen. But this solution determines the technology of device preparation. It has to be a thin-film technology. Thin-film transistors can be made of hydrogenated amorphous (nanocrystalline or microcrystalline) silicon, small-molecule or polymer organic semiconductors or some polycrystalline compound semiconductors (CdSe, for example). Hydrogenated amorphous silicon thin-film transistors have advantages

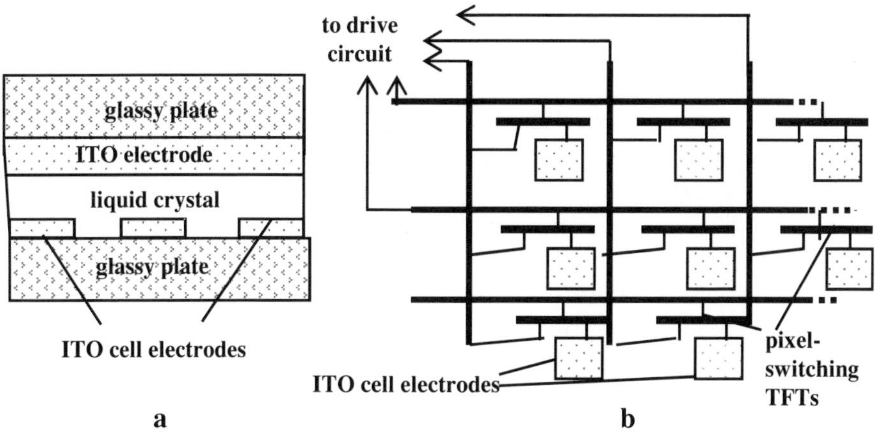

Figure 7.26. Structure (a) and diagram (b) of liquid crystal display (TFTs are thin-film transistors and ITO is SnO_2+In_2O_3).

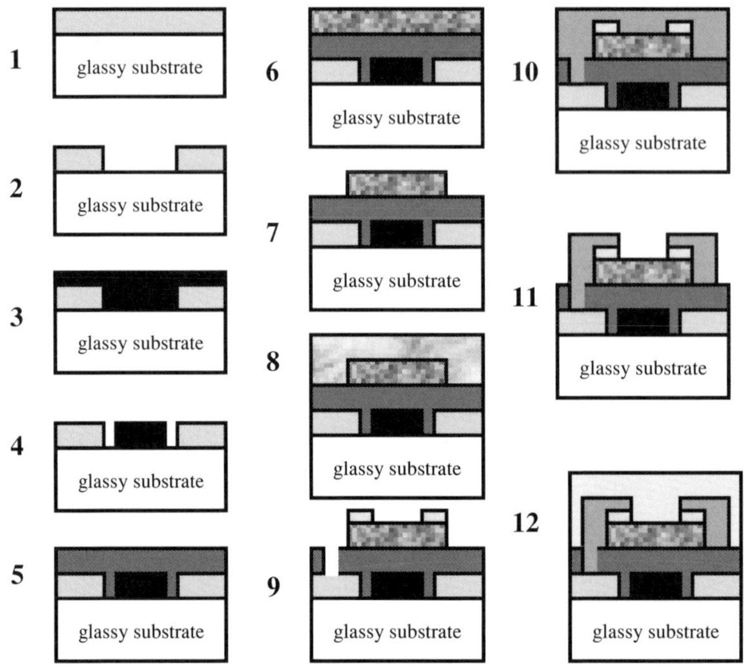

Figure 7.27. Basic steps of active matrix LCD fabrication (see text).

both in terms of technological base, and in terms of their characteristics. Since the deposition temperature of hydrogenated amorphous silicon films is in the range of 250–350°C (see Sec. 5.2) thin-film transistors can be prepared on a glass substrate. Therefore, these transistors are widely used both as pixel-switching devices in active matrix liquid crystal displays (LCD) and as peripheral driver circuits. The electrical circuit of the active matrix LCD is shown in Fig. 7.26(b).

Technological manufacturing process of the active matrix LCD includes the following operations (Fig. 7.27):

(1) Deposition of transparent ITO film on a glassy substrate.
(2) The first photolithography process to form ITO cell electrodes.
(3) Deposition of chromium film.
(4) The second photolithography process to form chromium gate electrodes and matrix horizontal buses.
(5) Deposition of silicon nitride (Si_3N_4) film.
(6) Deposition of undoped hydrogenated amorphous silicon film.
(7) The third photolithography process to form transistor active regions.
(8) Deposition of n^+-type hydrogenated amorphous silicon film.
(9) The fourth photolithography process to form transistor source and drain regions and windows on the silicon nitride.
(10) Deposition of aluminum film.

(11) The fifth photolithography process to form aluminum contacts to the drain and source of transistors and to cell ITO electrodes and to form matrix vertical buses.
(12) Deposition of protective layer.
(13) The sixth photolithography process to form windows on protective layer to create the contacts to matrix buses (not shown in the figure).

Thin-film transistors based on hydrogenated amorphous silicon have the following distinctive features:

- Current in the off-state is very small (about 10^{-12} A) due to the very high resistance of undoped hydrogenated amorphous silicon (see Sec. 4.1). In the on-state current increases by 6–8 orders of magnitude. This provides large on-off ratio (more than six orders of magnitude).
- All fabrication processes are carried out by conventional photolithographic technique that provides reproducible and cheap production of large scale integrated liquid crystal displays.
- Presence of localized states in the band gap of hydrogenated amorphous silicon limits the field-effect mobility of charge carriers in the transistor channel ($\mu_n = 0.5 - 1.0\,\text{cm}^2\,V^{-1}\text{s}^{-1}$ and $\mu_p \approx 0.01\,\text{cm}^2\,V^{-1}\text{s}^{-1}$). This leads to limitation of the switching speed.

There are two ways to increase the operation speed of transistors, namely the reduction of the channel length between source and drain, or increasing the field-effect mobility of the material. However, reduction of the channel length means decreasing minimum feature size of photolithographic process and, consequently, a rise in probability of defects. This is a serious problem in the case of active matrix LCD, since any technological defect causes visual defect. Therefore, reducing the channel length of transistor by traditional methods is not an acceptable way to solve the problem.

One possible solution of the problem noted above is a thin-film transistor with a vertical structure. Channel length of this transistor is determined by the thickness of the dielectric film and is equal to 0.5–1.0 μm (Aivasov *et al.*, 1995). The structure of the transistor is shown in Fig. 7.28. Films of metal, amorphous (or

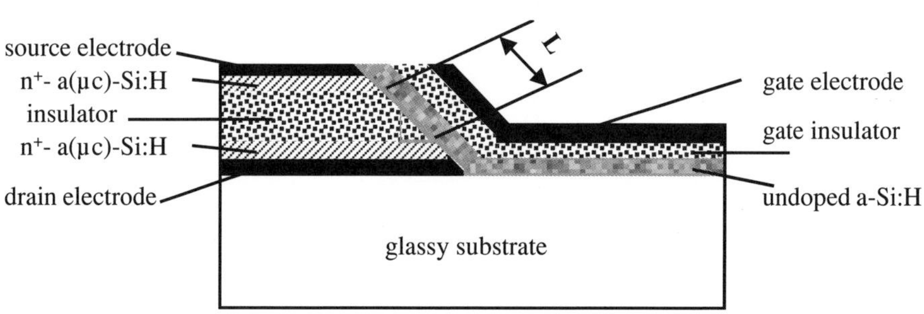

Figure 7.28. Thin-film transistor with vertical structure (L is channel length).

microcrystalline) hydrogenated n^+-type silicon, silicon nitride (thickness of about 1 μm), amorphous (or microcrystalline) hydrogenated n^+-type silicon and metal are deposited onto the substrate sequentially. Then photolithography and etching of films is carried out. The upper and lower metal films are the source and drain electrodes. Films of amorphous (or microcrystalline) hydrogenated n^+-type silicon are source and drain regions. Then films of undoped hydrogenated amorphous silicon, silicon nitride (gate insulator) and metal (gate electrode) are deposited. The channel is formed in the undoped a–Si:H film between two n^+-type layers, and its length (L) is approximately equal to the thickness of the silicon nitride layer. This allows to increase the operation speed of transistors by more than an order of magnitude at the same technological minimum features size.

Another way to increase transistor operation speed is to obtain a semiconductor material with a higher mobility of charge carriers. Increased mobility of electrons and holes in silicon thin-film transistors can be achieved by replacing the amorphous to nanocrystalline or microcrystalline hydrogenated silicon (Wagner *et al.*, 2003). The mobility of charge carriers increases with the size of crystal grains in the material (Fig. 7.29).

The nanocrystalline or microcrystalline silicon film can be obtained either by crystallization of initial a–Si:H film or from direct deposition. As a result thin-film n-type channel transistors with mobility of several tens of cm^2 V^{-1}s^{-1} (Matsumara *et al.*, 2003) and even with the mobility above 100 cm^2 V^{-1}s^{-1} (Stannowski, Rath and Schropp, 2003) were manufactured. Moreover, thin film p-type channel transistors have been created. This cannot be achieved in amorphous silicon due to low hole mobility.

The next goal of the flat panel industry is displays fabricated on flexible plastic foil substrates. Flexible displays are attractive due to light weight and ruggedness.

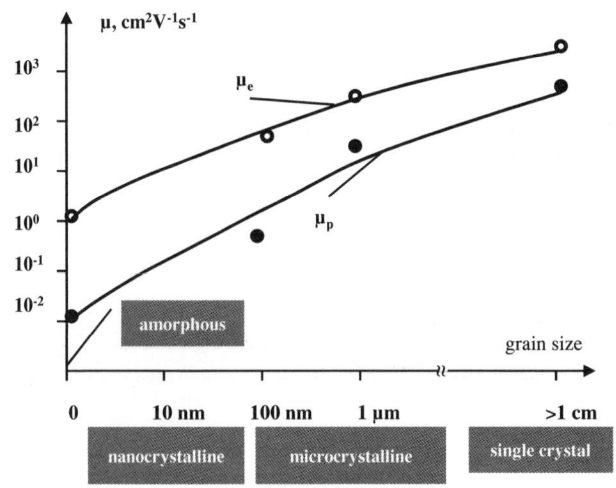

Figure 7.29. Dependence of electron (μ_e) and hole (μ_p) mobility in silicon on crystal grain size.

There are desirable features for portable applications today and for large area displays, electronic textiles and the mechatronic tomorrow (Wagner *et al.*, 2003).

However, plastic foils may be processed up to only 150–200°C. On the other hand, the optimal range of substrate temperature for device-quality hydrogenated amorphous silicon deposited by glow discharge decomposition method is 250–350°C. Decreasing deposition temperature leads to deterioration of a–Si:H film quality (see Sec. 5.2). To overcome this problem dilution of silane with hydrogen in the glow discharge decomposition method (Wagner *et al.*, 2003) or the method of hot wire chemical vapor deposition (see Sec. 5.4) are used. This allows to obtain device quality films of hydrogenated silicon at a lower substrate temperature. Example of the a–Si:H thin-film transistor structure made on SiN_x passivated polyimide foil is shown in Fig. 7.30 (Wagner *et al*, 2003).

For some applications (for example, in certain types of electrophotographic printers) thin-film transistors with a high (hundreds of volts) voltage source-drain and the small (a few volts) gate voltage are needed. High voltage thin film transistor based on hydrogenated amorphous silicon (Fig. 7.31(a)) has been created to solve this problem (Hack and Shur, 1987). As one can see from the figure, the gate electrode is located under the source area, and the inter-electrode space source — drain is large enough (about 20 μm). The part of the device that is located

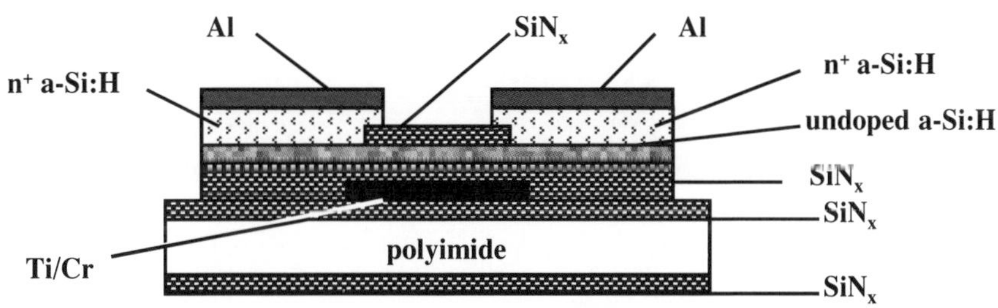

Figure 7.30. Cross section through a - Si:H thin film transistor made on SiNx passivated polyimide foil.

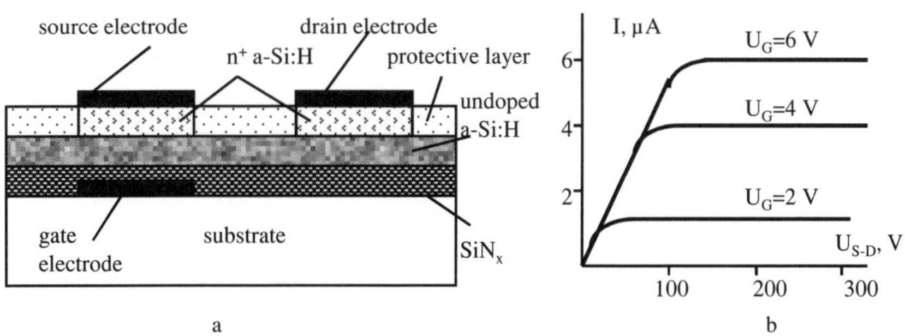

Figure 7.31. Structure (a) and current-voltage characteristics of the high voltage thin film transistor.

above the gate electrode operates as an ordinary field-effect transistor. Electrons from this region are injected into the inter-electrode source — drain space and the most part of the applied voltage drops there. Current-voltage characteristics of the transistor are shown in Fig. 7.31(b). Operating voltage of the transistor reaches 400 V at gate voltages from 0–10 V.

Conclusion

In conclusion I would like to return to the question posed at the beginning of this book: "Is the attention to disordered materials a step forward or a step back from the point of view of solid state physics and microelectronics?" I hope that after reading this book the reader will agree with me that physics and applications of disordered semiconductors is not only a step forward but a step into the future. There is a number of currently unsolved problems in the physics of disordered semiconductors and their applications that offer a wide and interesting field of activity for young physicists and designers of various devices in microelectronics, nanoelectronics, optoelectronics and other fields of technology. I wish you every success in this exciting area of study and research.

Anatoly Popov

Disordered Semiconductors: Physics and Applications by Anatoly Popov

www.panstanford.com
978-981-4241-76-2

References

Adler D., Density of states in the gap of tetrahedrally bonded amorphous semiconductors, *Phys. Rev. Lett.*, **41**, 1755–1758 (1978).

Aitken B. G., Structure-property relationships in Ge — As selenide and sulphide glasses, Proceedings International Congress on Glass, Edinburgh, Invited Papers, pp.135–140 (2001).

Aivasov A. A., Budagyn B. G., Vihrov S. P. and Popov A. I., Disordered semiconductors, Moscow: "Vishaya shcola", p.352 (in Russian), (1995).

Anderson P. W., Model for the electronic structure of amorphous semiconductors, *Phys. Rew. Lett.*, **34**, 953–955 (1975).

Andrievsky A. I. and Nabitovich I. D., Structure of β-cubic modification of selenium, *Sov. Phys. Crystallogr.*, **5**, 442-443 (1960).

Andrievsky A. I., Nabitovich I. D. and Krinykovicn P. K., Structure of selenium in thin layers, *Sov. Phys. Dokl.*, **4**, 16–18 (1959).

Averynov V. L., Kolobov A. V., Kolomiets B. T., Lyubin V. M. and Fedorov V. A., Photo-darkening and photo-bleaching in chalcogenide glassy semiconductor films As-Se and As-S systems, Physical Phenomena in Non-Crystalline Semiconductors, Proc. Conf. "Amorphous Semiconductors-80", Kishinev, AN SSSR, pp.213–215 (in Russian),(1980).

Baratov A. G. and Popov A. I., Molecular structure and properties of chalcogens (sulphur, selenium and tellurium), Erevan: "Aiastan", 88 p (in Russian),(1990).

Bowman D. L. and Schottmiller J. C., Vitreous Bi – Se layers as near-infrared photodetectors, *J. Appl. Phys.*, **39**, 1659–1662 (1968).

Carlson C., Electrophotography, U.S. Patent 2.297.691,(1942).

Carlson D., Solar energy conversion, In: The Physics of Hydrogenated Amorphous Silicon I. Structure, Preparation and Devices, ed. by J. Joannopoulos and G. Lucovsky, Springer-Verlag, pp.203–244,(1984).

Cherkasov Yu. A. and Kreitor L. G., Exiton absorption and photoconductivity in amorphous analogs of various crystalline forms of selenium, Fisica tverdogo tela, **16**, 2407–2410 (in Russian), (1974).

Cohen M. H., Fritzsche H. and Ovshinsky S. R., Simple band model for amorphous semiconducting alloys, *Phys. Rev. Lett.*, **22**, 1065–1068 (1969).

Cormack A. N., Du J. and Zeitler T, Molecular Modeling of Glass, Proceedings of the International Congress of Glass, Edinburg, Invited papers, pp. 170–174 (2001).

Davis E. A. and Greaves G. N., Amorphous arsenic — properties and structural model, Proceedings of the International Conference on Amorphous and Liquid Semiconductors, *Leningrad*, pp.212–220 (1976).

Davis E. A. and Mott N. F., Condition in non-crystalline systems, *Phil. Mag.*, **22**, 903–922 (1970).

Dembovsky S. A., Empirical glass forming theory in chalcogenide systems, *J. neorganicheskoy Chimie*, **22**, 3187–3199 (in Russian) (1977).

Dembovsky S. A. and Chechetkina E. A., The molecular-kinetic model for critical cooling rate of glass forming melts, *Mat. Res. Bul.* **16**, 723–727 (1981a).

Dembovsky S. A. and Chechetkina E. A., Kinetico-thermodynamic aspect of glass formation and critical cooling rates in chalcogenide systems, *Mat. Res. Bul.* **16**, 505–511 (1981b).

Dembovsky S. A., Chechetkina E. A. and Kozukhin S. A., Anomalous influence of weak magnetic fields on diamagnetic glassy semiconductors, *JETP Lett.*, **41**, 88–90 (1985).

Dembovsky S. A., Zyubin A. S. and Grigor'ev F. V., Modeling of hypervalent configurations, valence alternation pairs, deformed structure and properties of a – S and a – As_2S_s, *Semiconductors*, **32**, 899–916 (1998).

Elliot S. R., Medium-range order in amorphous materials: documented cases, *J. Non-Cryst. Solids*, **97** and **98**, 159–162 (1987).

Elliott S. R. and Davis E. A., Defect states in Group-V amorphous semiconductors, *J. Phys. C*, **12**, 2577–2587 (1979).

Flasck R, Izu M, Supru K, Anderson T, Ovshinsky S. R. and Fritzsche H., Optical and electronic properties of modified amorphous materials, Amorphous and Liquid Semiconductors, *Proc. of the 7th Intern. Conf. on Amorphous and Liquid Semiconductors*, ed. by W. E. Spear, Edinburgh, 524–528,(1977).

Greaves G. N., Elliott S. R. and Davis E. A., Amorphous arsenic, *Adv. Phys.*, **28**, 49–141 (1979).

Gregg A, Blieden R, Chang A. and Ng H., Performance analysis of large scale, amorphous silicon, photovoltaic power system. In: Photovoltaic Specialists Conference, 2005, Conference Record of the Thirty-first IEEE, pp. 1615–1618 (2005).

Grigorovici R., Structure of amorphous semiconductors, in: Amorphous and Liquid Semiconductors, ed. by J. Tauc, Plenum Press, pp. 45–99 (1974).

Golikova O. A., Doping and pseudodoping of amorphous silicon, *Sov. Phys. Semiconductors*, **25** pp. 915–938 (1991).

Gubanov A. I., Quantum-electronics theory of amorphous semiconductors, Moscow-Leningrad: Academy of Science (in Russian), p. 250 (1991).

Gurevich S. B., Ilyshenko N. N., Kolomiets B. T., Lyubin V. M., Fedorov V. A. and Shilo V. P., Holographic media on the base of chalcogenide glassy semiconductor films of the As - Se system. In: Holography and information transmission, Leningrad, Nauka, pp. 146–157 (in Russian), (1976).

Gurevich S. B., Ilyshenko N. N., Kolomiets B. T., Lyubin V. M. and Shilo V. P., Photo-induced changes of optical properties of some chalcogenide glassy semiconductors, in: Structure and properties of non-crystalline semiconductors, *Proc. 6th Intern. Conf. on amorphous and liquid semiconductors*, FTI, Leningrad, pp. 451–455 (in Russian) (1976b).

Hack M. and Shur M., Analysis of amorphous silicon thin-film transistors, *J. Non-Cryst. Solids*, **97&98**, 1291–1294 (1987).

Juska G, Arlauskas K. and Montrimas E., Features of carriers at very high electric fields in a–Se and a–Si : H, J. Non-Cryst. *Solids*, **97&98**, 559–562 (1987).

Kaplan D., CVD material, In: The Physics of Hydrogenated Amorphous Silicon I. Structure, Preparation and Devices, ed. by J. Joannopoulos and G. Lucovsky, *Springer-Verlag* pp. 177–202 (1984).

Kastner M., Radiative and non-radiative processes at valence alternation and other defects in lone-pair semiconductors, Proc. of VII intern. conf. of amorphous and liquid semiconductors, Edinburgh, pp. 504–508 (1977).

Kastner M, Adler D. and Fritzsche H., Valence-alternation model for localized gap states in lone-pair semiconductors, *Phys. Rew. Lett.*, **37**, 1504–1507 (1976).

Kastner, M. and Hudgens, S. L. Evidence for neutrality of luminescence centers in chalcogenide glasses, *Phil. Mag.*, *B* **37**, 665–681 (1978).

Kawamura, T., Yamamoto, N. and Nakayama, Y Electrophotographic application of amorphous semiconductors, Amorphous semiconductor technologies and devices, *Jap. Annual Rev. in Electronics, Computers and Telecommunications*, **6**, 325–336 (1983).

Keating, P. N. Effect of invariance requirements on the elastic strain energy of crystals with application to the diamond structure, *Phys. Rev.*, **145**, 637–645 (1966).

Khokhlov, A. F., Mashin, A. I. and Khokhlov, D. A. New allotropic form of silicon, *JETF Lett.*, **67**, 675–679 (1998).

Knights, J. S. and Lujan, R. A. Microstructure of plasma-deposited a – Si:H films, *Appl. Phys. Lett.*, **35**, 214–216 (1979).

Kolobov, A. V. and Shimakava, K. Photoinduced metastability in CGS, In: Electronic phenomena in chalcogenide glassy semiconductors, ed. by K. Tsendin, St.-Peterburg, "Nauka", pp. 365–395 (in Russian) (1996).

Kolobov, A. V., Fons, P., Frenkel, A. I., Ankudinov, A. L., Tominaga, J. and Uruga, T. Understanding of phase-change mechanism of rewritable optical media, *Nature materials,* **3**, 703–708 (2004).

Kolomietz, B. T., Glassy state (Stekloobraznoe sostoynie), Moscow — Leningrad (in Russian) (1960).

Kolomiets, B. T., Lebedev, E. a. and Taksami, I. A. Mechanism of the breakdown in films of glassy chalcogenide semiconductors, *Sov. Phys. Semicond.*, **3**, 267–269 (1969).

Kolomietz, B. T. Studies of chalcogenide vitreous semiconductors in Ioffe physico-technical institute, Journal de Physique, Colloque C4, **42**, C4-887–C4-897 (1981).

Kolomietz, B. T., Lyubin, V. M. and Shilo, V. P. Photo-induced changes in the etching rate of chalcogenide glasses, *Physics and Chemistry of Glasses,* **4**, 351–357 (in Russian) (1978).

Kostylev, C. A. and Shkut, V. A. Electronic switching in amorphous semiconductors, Kiev, "Naukova dumka", p. 203 (in Russian) (1978).

Kostyshin, M. T., Mikhaylovskaya, Ye. V. and Romanenko, P. F. On the effect on photosensitivity of thin semiconductor layers on metallic substrates, *Soviet Physics Solid State,* **8**, 451–452 (1966).

Kurashige, M., Tanioka, K., Ooka, M. Hirai, T. 1-inch all electrostatic super HARD tube and camera for HDTV – a new possibility for very dark images with super-fine resolution, *Proc. SPEI,* **1987**, pp. 1–10 (1993).

Kuwano, Y. and Ohnishi, M. Integrated amorphous silicon solar modules, In: Japan annual reviews in electronics, computers and telecommunications, Amorphous semiconductors: technology and devices, ed. by Y. Hamakawa, OHM, Norton-Holland, pp. 204–227 (1983).

Lacaita, A. L. and Wounters, D. J., Phase-change memories, *Phys. Stat. Sol.* (a), **205**, 2281–2297 (2008).

Lebedev, A. A., On polymorphism and annealing in glass, *Proceedings of State Optical Institute,* St.-Peterburg, **2(10)**, 1–21 (in Russian) (1921).

Long, M., Galison, P. and Alben, R. Structural model of amorphous selenium and tellutium, *Phys. Rev.*, **B13**, 1821–1829 (1976).

Lowrey, T. A., Hudgens, S. J., Czubatyj, W., Dennison, C. H., Kostylev, S. A. and Wicker, G. C. Characteristics of OUN phase change materials and devices for high density nonvolatile commodity and embedded memory applications, *Mat. Res. Soc. Symp. Proc.* **803**, HH2.1.1–2.1.12 (2004).

Lukovsky, G. Specification of medium-range order in amorphous materials, *J. Non-Cryst. Solids,* **97&98**, pp. 155–158 (1987).

Lukovsky, G. and Galeener, F. L. Intermediate range order in amorphous solids, *J. Non-cryst. Solids,* **35&36**, 1209–1214 (1980).

Madan, A. Devices using fluorinated material, In: The Physics of Hydrogenated Amorphous Silicon I. Structure, Preparation and Devices, ed. by J. Joannopoulos and G. Lucovsky, *Springer-Verlag,* pp. 245–283 (1984).

Malaurent, J. C. and Dixmier, J. A random chain model for amorphous selenium, The structure of non-crystalline materials, Cambridge University Press, pp. 49–51 (1977).

Maruyama, M. Development of the underwater super-HARD color TV camera, *Proc. ITEC'92, N* **22-6**, pp. 425–426 (1992).

Mashin, A. I. and Khokhlov, A. F. Multiple bonds in hydrogen-free amorphous silicon, *Semiconductors*, **33**, 911–914 (1999).

Masters, J. I. and Goldberg, G. M. Process of doping silver image in chalcogenide layer, US Patent 4,269,935, Class 430/323 (1981).

Matsumura, H., Umemoto, H., Izumi, A. and Masuda A. Recent progress of Cat-CVD research in Japan – bridging between the first and second Cat-CVD conferences, *Thin Solid Films*, **430**, 7–14 (2003).

Minaev, V. S. Glassforming semiconductor alloys, Moscow: "Metalurgija" (in Russian), p. 406 (1991).

Miyauchi, K., Qui, J., Shojiya, M. and Kawamura, N. Structural study of GeS_2 glasses permanently densified under high pressures up to 9 GPa, *J. Non-Cryst. Solids*, **279**, 186–195 (2001).

Mizushima, Y. and Yoshikawa, A. Photoprocessing and lithographic applications, Japan *Annual Reviews in Electronics, Computers and Telecommunications*, **2**, 277–295 (1982).

Moller, S., Perlov, C., Jakson, W., Taussing, C. and Forrest, S. R. A polymer/semiconductor write-once-read-many-times memory, *Nature*, **426**, 161–169 (2003).

Morigaki, K. Physics of amorphous semiconductors, Imperial College Press and World Scientific Publishing, p.418 (1999).

Mott, N. F. and Davis, E. A. Electron process in non-crystalline materials, Clarendon Press, Oxford (1979).

Okamoto, H., Hattori, K. and Hamakawa, Y. Hall effect near the mobility edge, *J. Non-Cryst. Solids*, **164–166**, 445–448 (1993).

Ovshinsky, S. R. Reversible electrical switching phenomena in disordered structures, *Phys. Rew. Lett.*, **21**, 1450–1453 (1968).

Ovshinsky, S. R. Chemistry and structure in amorphous materials: the shapes of thinks to came, in: Disordered materials. Science and technology. Selected papers, Plenum Press, pp. 333 – 342 (1991).

Ovshinsky, S. R. Chemical modification of amorphous chalcogenides, Amorphous and Liquid Semiconductors, Proc. of the 7th Intern. Conf. on Amorphous and Liquid Semiconductors, ed. by W. E. Spear, Edinburgh, pp. 519–523 (1977).

Park, W.-D. and Tanioka, K. Avalanche-type high sensitive image pickup tube using an a-Se photoconductive target, Jpn. *J. Appl. Phys.*, **42** L209–L211 (2003).

Pazin, A. V. and Borisova, Z. U. Glass-forming region in bismuth-germanium-selenium system, Vestnik LGU, Phisika i himia, **4**, 22, 140–144 (in Russian) (1969).

Petkov, K. and Fedorov, B. A. Photo-induced changes of chalcogenide glassy semiconductor films As-S system, Proc. Intern. Conf. "Amorphous semiconductors – 82", Bucharest, CIP-AP, 205–207 (in Russian) (1982).

Pfister, G. and Scher, H. Dispersive (non-Gaussian) transient transport in disordered solids, *Adv. Phys.*, **27**, 747–798 (1978).

Phillips, J. C. Topology of covalent non-crystalline solids I: short-range order in chalcogenide alloys, *J. Non-Cryst. Solids*, **34**, 153–181 (1979a).

Phillips, J. C. Hollow claster models of atomic structure of chalcogenide glasses, *Inst. Phys. Conf. Ser.*, **43**, 705–708 (ch. 21) (1979b).

Phillips, J. C. Chemical bonding, internal serfaces, and the topology of non-crystalline solids, *Phys. Status Solidi*, **B101**, 473–479 (1980).

Pimentel, G. C. and Spratley, R. D. Chemical bonding clarified through quantum mechanics, Holden-Day Inc (1970).

Polk, D. E. Structural model for amorphous silicon and germanium, *J. Non-Cryst. Solids,* **5**, 365–376 (1971).

Poltavzev, Yu. G. Structure of semiconductor melts, Metallulgiya, Moscow (in Russian) (1984).

Popov, A. I. The influence of the structural changes on viscosity of selenium, Fisika i Khimia Stecla, **6**, 307–311 (in Russian) (1980).

Popov, A. I. Determination of the rigidity threshold of the structural network in covalent glasses, *Glass Phys. Chem.,* **20**, 553–554 (1994).

Popov, A. I. Electro-photographic materials, Materials Sciences Transactions (Materialovedenie) (in Russian), **2**, 27–33 (1997).

Popov, A. I. Structural characteristics and structural modifications of non-crystalline semiconductors, *J. Optoelectronics and Advanced Materials,* **4**, 481–488 (2002).

Popov, A. I. Atomic structure and structural modification of glass, *Semiconductors and Semimetals,* **78**, 51–95 (2004).

Popov, A. I. Physics and technology of disordered semiconductors, Publishing house MPEI, Moscow, (in Russian) p. 272 (2008).

Popov, A. I. Two generation of phase-change memory devices: differences and common problems, *Phys. Status Solidi (b),* **246**, 1837–1840 (2009).

Popov, A. I., Ligachev, V. A., Vasil'eva, N. D. and Stuokacn, S. N. Preparation of thin hydrogenated carbon films by HF sputtering method, Proceedings of Ukrainian Vacuum Society, Kiev, **1**, 248–251 (in Russian) (1995) .

Popov, A. I., Michalev, N. I. and Shemetova, V. K. Structural modification of some glassy chalcogenudes, *Philos. Mag., B* **47**, 73–81 (1983).

Popov, A., Mikhalev, N., Karalynets, A. and Vasilyeva N. Photosensitivity of selenium — bismuth films with varigap structure, *Proc. Int. Conf. Thin films Physics and Application,* Shanghai, China SPIE, **1519**, 457–462 (1991).

Popov, A. I., Vorontsov, V. A. and Popov, I. A. Levels of structural modification of noncrystalline semiconductors and limits of their applicability, *Semiconductors,* **35**, 637–642 (2001).

Popov, A. I., Savinov, I. S. and Voronkov, E. N. Simulation of phase-change processes in non-volatile memory cells, *J. Non-Cryst. Solids,* **352**, 1624–1627 (2006).

Popov, I. A. Structural modification of amorphous hydrogenated silicon with help of ultraviolet radiation, *Semiconductors,* **30**, 258–265 (1996).

Popov, N. A. New model of defects in chalcogenide glassy semiconductors, *Pisma v GETF,* (in Russian) **31**, 437–440 (1980).

Popov, N. A. Quasimolecular defects in glassy chalcogenide semiconductors, *Soviet Phys. Semicond.,* **15**, 212–215 (1981).

Resnik, A., Lui, B. J. M., Lubin, V., Klebanov, M., Ohkawa, Y., Matsubara, T., Miyakawa, K., Kubota, M., Tanioka, K., Kawai, T. and Rowlands, J. A. The effect of temperature on photoinduced metastability in avalanche a-Se layers, *J. Non-Cryst. Solids,* **352**, 1595–1598 (2006).

Richer, y. and Breiting, G. Verschiedene formen von amorphen selen, Z. Naturforsch, **26a**, 1699–1708 (in German) (1971).

Tanaka, K. Reversible photostructural change: mechanism, properties and application, *J. Non-Cryst. Solids,* **35&36**, 1023–1034 (1980).

Tanaka, K. Photo-induced phenomena in amorphous semiconductors, in: Fundamental physics of amorphous semiconductors, ed. by f. Yonezawa, Berlin, Springer-Verlag, pp. 104–118 (1981).

Sadoc, J. F., Mosseri, R. Order and disorder in amorphous tetrahedrally coordinated semiconductors: a curved-space description, *Phil. Mag.*, **B 45** 467–483 (1982).

Schottmiller, J. C., Bowman, D. L. and Wood, C. *J. Appl. Phys.* **39**, 1663–1669 (1968).

Sherchenkov, A. A. and Budagyn, B. G. Physics and technology of semiconductor power converters, part 2, MIET, Moscow, 280 p. (in Russian) (2007).

Sladkov, A. M., Korshak, V. V., Kudryavtsev, Yu. P. and Kasatochkin, V. I. USSR Discoverer's Certificate No 107, Byull. Izobret., **6** (3) (1972).

Smorgonskaya, E. A. and Tsendin, K. D. Atomic and electronic structure of chalcogenide glassy semiconductors. In: Electronic Phenomena in Cnalcogenide Glassy Semiconductors, "Nauka", Sant-Peterburg, pp. 9-33 (in Russian) (1996).

Spear, W. E. and Le Comber, P. G. Substitutional doping of amorphous silicon, *Solid State Commun.*, **17**, 1193–1196 (1975).

Spear, W. E. and Le Comber, P. G. Electronic properties of substitutionally doped amorphous Si and Ge, *Philos. Mag.*, **33**, 935–949 (1976).

Spear, W. E. and Le Comber, P. G. Fundamental and applied work on glow discharge material, In: Topics in Applied Physics, The physics of hydrogenated amorphous silicon I, Springer, **55**, pp. 63–118 (1984).

Staebler, D. L. and Wronski, C. R. Reversible conductivity changes in discharge-produced amorphous Si, *Appl. Phys. Lett.* **31**, 292–294 (1977).

Stannowski, B., Rath, J. K. and Schropp, R. E. I., Thin-film transistors deposited by hot-wire chemical vapor deposition, *Thin Solid Films*, **430**, 220–225 (2003).

Street, R. A. Non-radiative recombination, photostructural changes and Urbach's rule in chalcogenides, Proc. of VII intern. conf. of amorphous and liquid semiconductors, Edinburgh, 509–513 (1977).

Street, R. A. and Mott, N. F. States in the gap in glassy semiconductors, *Phys. Rew. Lett.*, **35**, 1293–1296 (1975).

Stutzmann, M., Biegelsen, D. K. and Street, R. A. Datailed investigation of doping in hydrogenated amorphous silicon and germanium, *Phys. Rev.* **B 35**, 5666–5701 (1987).

Sunayama, H., Yamada, K., Karasawa, M. and Ishibashi, K. Preparation of poly-Si films by Cat-CVD for thin transistor, *Thin Solid Films*, **430**, 226–229 (2003).

Thompson, M. Sputtered material, In: The Physics of Hydrogenated Amorphous Silicon I. Structure, Preparation and Devices, ed. by J. Joannopoulos and G. Lucovsky, Springer-Verlag pp. 119 – 175 (1984).

Vasil'eva, N. D. and Popov, A. I. Effect of deposition conditions on the atomic structure of hydrogenated amorphous carbon films, *Inorg. Mater.*, **31**, 417–422 (1995).

Vasil'eva, N. D., Popov A. I. and Khalturin, V. P. Application of subgradient method to the non-crystalline semiconductor structure simulation, Proceedings of the International Conference "Amorphous Semiconductors – 82", Bucharest, 50–52 (in Russian) (1982).

Vikhrov, S. P., and Bodygin N. V. New approach to technology system design (preparation of a-Si:H layers), RSRTA, Ryzan, 108 (in Russian) (1994).

Voronkov, E. N., Popov, A. I., Savinov, I. S. and Fairushin, A. R. Percolation breakdown of amorphous semiconductors, *J. Non-Cryst. Solids*, **352**, 1578–1581 (2006) .

Voyles, P. M., Zotov, N., Nakhmanson, S. M., Drabold, D. A., Gibson, J. M., Treacy, M. M. J. and Keblinsky, P. Structural and physical properties of paracrystalline atomic models of amorphous silicon, *J. Appl. Phys.*, **90**, 4437–4451 (2001).

Wagner, S., Gleskova, H., Cheng, I-C. and Wu, M. Silicon for thin-film transistors, *Thin Solid Films*, **430**, 15–19 (2003).

Weiss, P. Superfast atom shuffling inspires data-storage alternatives, *Science News*, **167**, 363–364 (2005).

Wuttig, M. and Yamada, N. Phase-change materials for rewriteable data storage, Nature materials, **6**, 824–832 (2007).

Yamada, N. Phase-change material for use in rewritable dual-layer optical disk, *Proc. SPIE*, **4352**, pp. 55–63 (2002).

Yeung, F., Ahn, S-J., Hwang, Y-N., Jeong, C-W., Song, Y-J., Lee, S-J., Lee, S-H., Ryoo, K-C., Park, J-H., Shin, J-M., Jeong, W-C., Kim, Y-T., Koh, G-H., Jeong, G-T., Jeong, H-S. and Kim, K. $Ge_2Sb_2Te_5$ confined structures and integration of 64 Mb phase-change random access memory, Jap. *J. of Applied Physics*, **44**, pp. 2691–2695 (2005).

Yukimoto, Y Multi-layered and stacked solar cells, Amorphous semiconductor technologies and devices, Jap. *Annual Rev. in Electronics, Computers and Telecommunications*, **6**, 228–238 (1983).

Yun, S. S., Li, H., Cappelletti, R. L., Enzweiler, R. N. and Boolchand, P. Onset of rigidity in $Se_{1-x}Ge_x$ glasses: ultrasonic elastic module, *Phys. Rev.*, **B39**, pp. 8702–8706 (1989).

Zakery, A. and Elliott, S. R. Optical properties and applications of chalcogenide glasses: a review, *J. Non-Cryst. Solids*, **330**, pp. 1–12 (2003).

Color Inserts

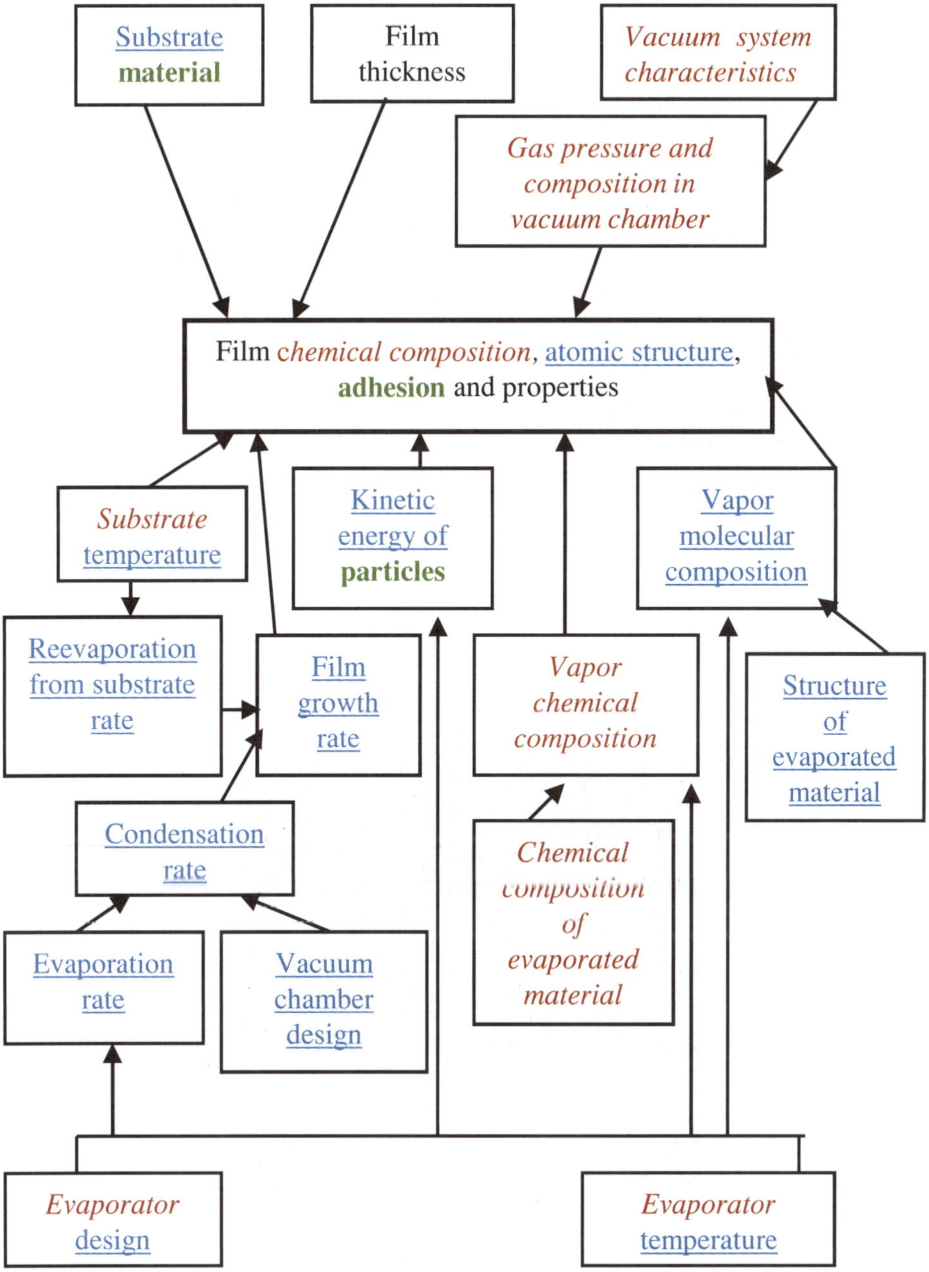

Figure 5.5. Determined film characteristic factors.

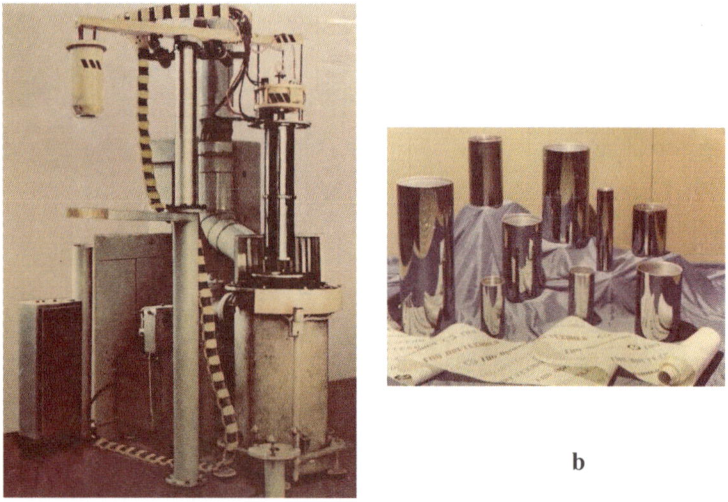

b

a

Figure 5.7. Electro-photographic drums manufacturing plant (a) and selenium drums (b).

Figure 6.13. Airscape taken at night with the CCD camera (top) and the Super-HARP camera (bottom) (M. Kurashige, unpublished).

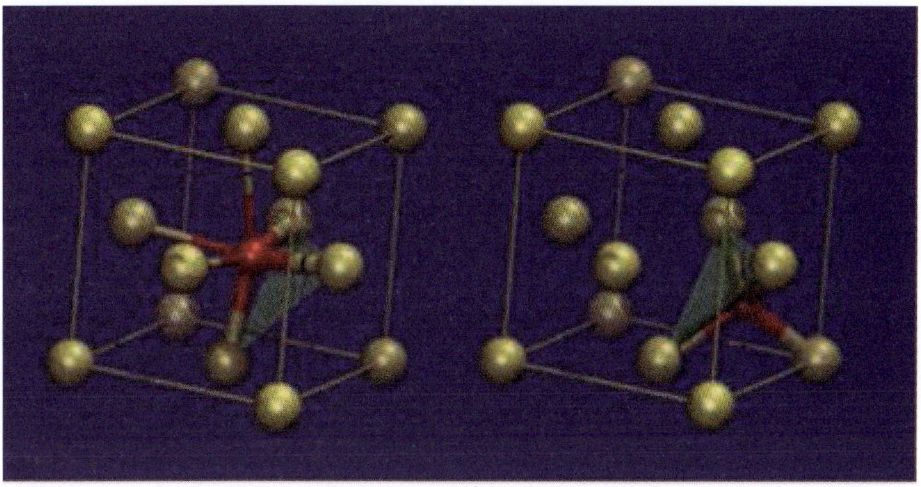

a b

Figure 6.27. Fragments of the local structure of GST225 around germanium atoms in the crystalline (a) and amorphous (b) states (stronger covalent bonds are shown as thicker lines whereas weaker bonds are shown as thinner lines) (Kolobov et al., 2004).

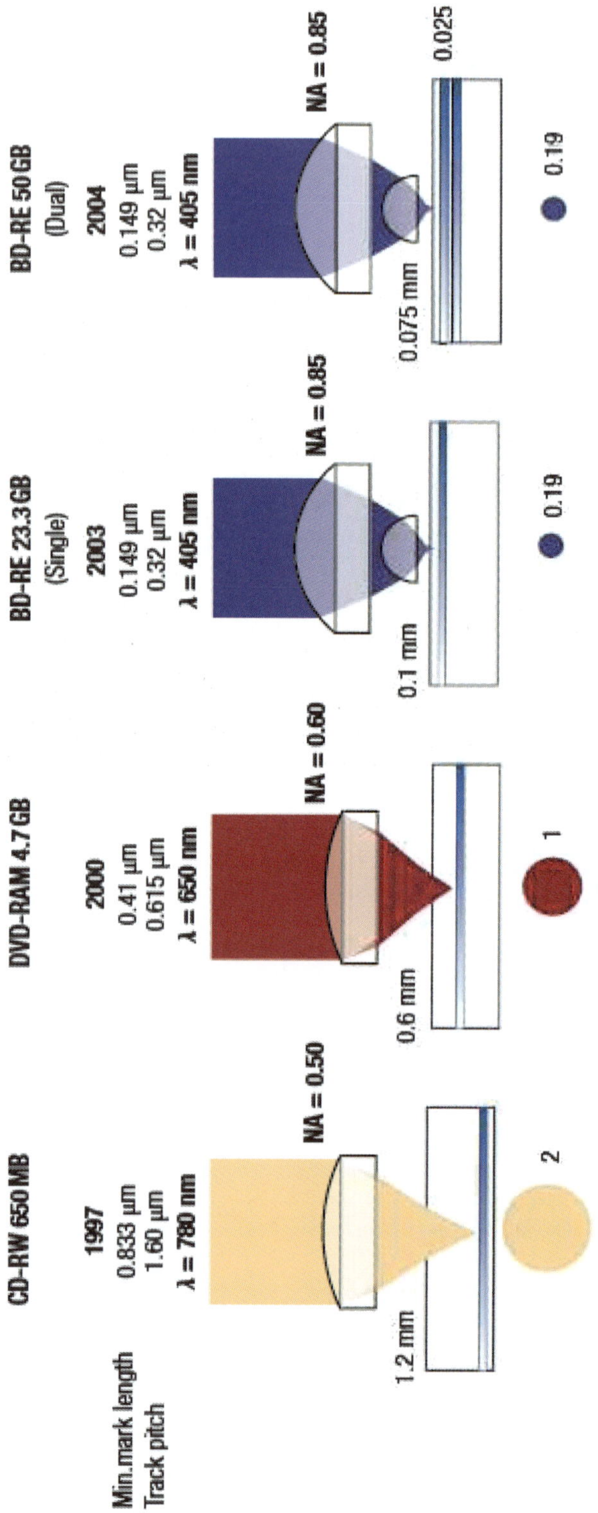

Figure 6.28. Comparison of Rewritable optical disc charactristics.

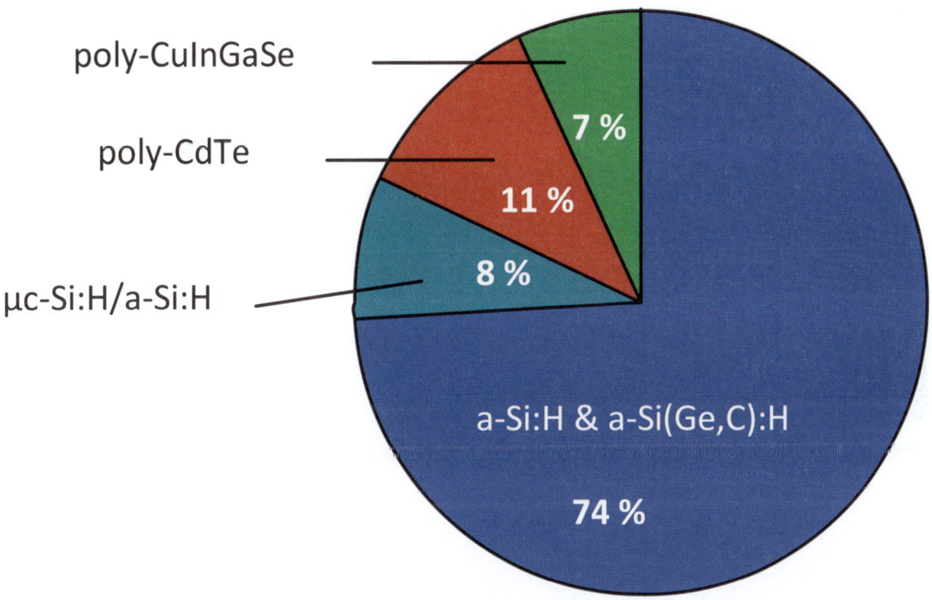

Figure 7.1. World production (sales) of thin-film solar sells.

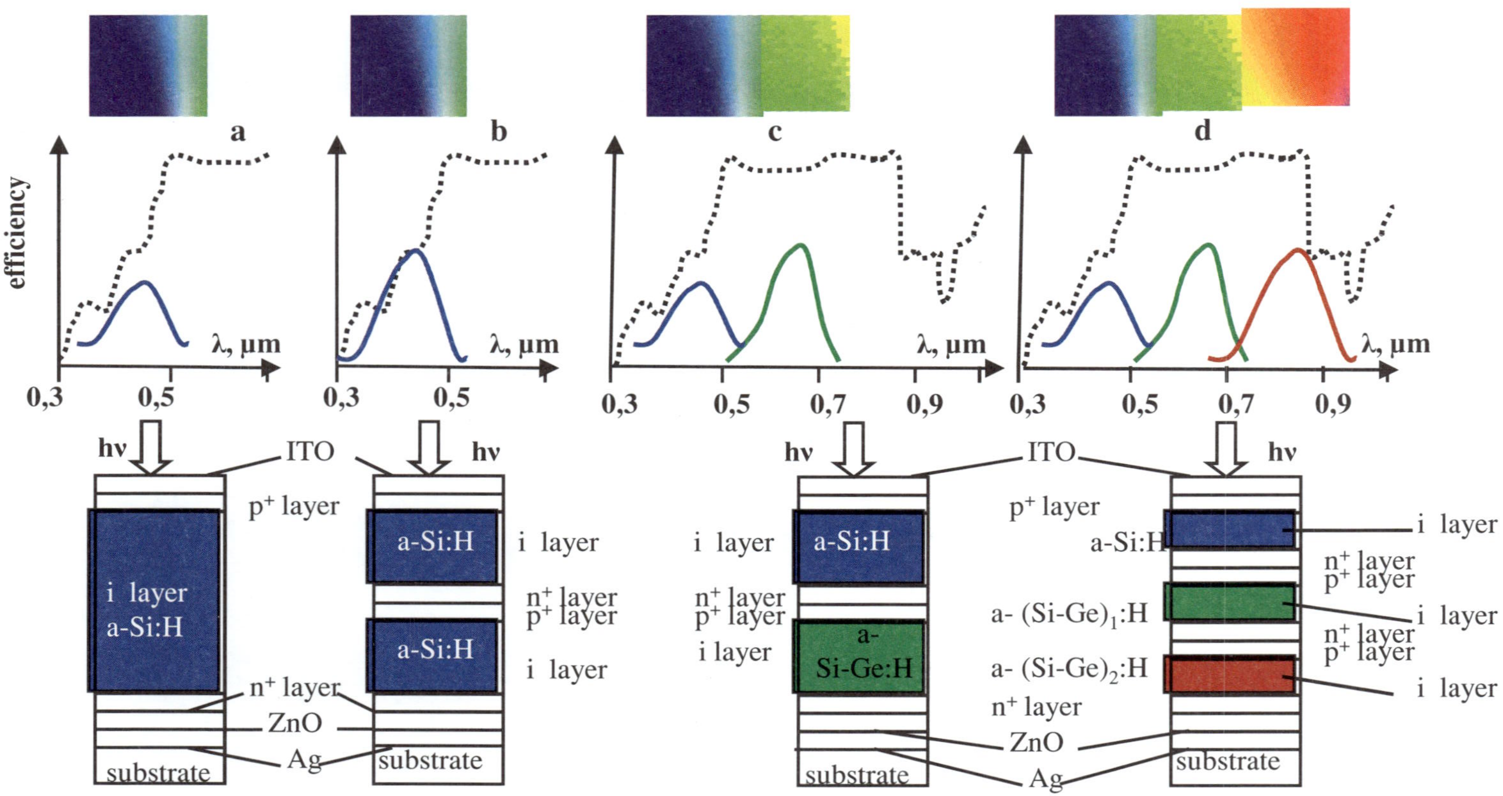

Figure 7.6. Schematic illustration of the spectral dependence of the efficiency(top) and structure(bottom) ofmultiple-junction solar cells: singlejunction (a) double-junction with the same gap (b) double-junction with dual gap (c) and trible junction (d) (the dashed line is the spectrum of sunlight on the Earths surface).

Figure 7.12. Uni-Solar amorphous silicon photovoltaic power system in California, USA (4800 laminates $5.4 \times 0.4\ m^2$, 500 kW, Gregg *et al.*, 2005).

Index